Inspire Science

Be a Scientist Notebook

Student Journal

Grade 5

Mc
Graw
Hill
Education

mheducation.com/prek-12

Mc Graw Hill Education

STEM McGraw-Hill is committed to providing instructional materials in Science, Technology, Engineering, and Mathematics (STEM) that give all students a solid foundation, one that prepares them for college and careers in the 21st century.

Send all inquiries to:
McGraw-Hill Education
8787 Orion Place
Columbus, OH 43240

ISBN: 978-0-07-678227-7
MHID: 0-07-678227-1

Printed in the United States of America.

10 11 12 13 LMN 23 22 21 20

Our mission is to provide educational resources that enable students to become the problem solvers of the 21st century and inspire them to explore careers within Science, Technology, Engineering, and Mathematics (STEM) related fields.

TABLE OF CONTENTS

Structure and Properties of Matter

Module Opener...2

Lesson 1   Matter's Structure................................4

Lesson 2   Matter's Properties..............................20

Lesson 3   Metals and Nonmetals.........................34

Module Wrap-Up..47

Physical and Chemical Changes

Module Opener...50

Lesson 1   Physical Changes.................................52

Lesson 2   Mixtures and Solutions........................66

Lesson 3   Chemical Changes...............................80

Module Wrap-Up..94

Plant and Animal Needs

Module Opener...96

Lesson 1   Plants and Photosynthesis....................98

Lesson 2   Animals and Cellular Respiration............110

Lesson 3   Plants and Cellular Respiration..............124

Module Wrap-Up..138

Matter in Ecosystems

Module Opener...140

Lesson 1   Interactions of Living Things.................142

Lesson 2   Balance in Ecosystems.........................156

Lesson 3   Cycles in Ecosystem............................172

Module Wrap-Up..186

HIRO
Ocean Engineer

TABLE OF CONTENTS

Interactions of Earth's Major Systems

Module Opener..**188**

Lesson 1   Earth's Major Systems..**190**

Lesson 2   Effects of the Geosphere...**202**

Lesson 3   Effects of the Hydrosphere....................................**218**

Lesson 4   Effects of the Atmosphere.....................................**234**

Lesson 5   Effects of the Biosphere...**248**

Module Wrap-Up...**262**

The Solar System and Beyond

Module Opener..**264**

Lesson 1   Movements of the Sun, Earth, and the Moon.....................**266**

Lesson 2   Patterns of the Moon..**282**

Lesson 3   Objects in Space..**296**

Lesson 4   Stars and Star Patterns..**310**

Module Wrap-Up...**322**

VKV Visual Kinesthetic Vocabulary

Introduction..**324**

Visual Kinesthetic Vocabulary Cut-Outs..................................**327**

Check out the activities in every lesson!

HIRO
Ocean Engineer

InspireScience

This is your own personal science journal where you will become scientists and engineers. Use this book to answer questions and solve real-world problems.

This is YOUR journal!
Personalize it!

MAYA
Geologist

Structure and Properties of Matter

Science in Our World

Barges can be used to transport large quantities of goods from one place to another. Look at the photo of the large barges and other boats floating on the open water. What questions do you have about these boats?

_____

_____

_____

abc Key Vocabulary

Look and listen for these words as you learn about the structure and properties of matter.

alloy	atom	buoyancy
compound	density	element
mass	matter	metal
molecule	nonmetal	volume

How can I use what I know about different types of metals to identify how they can be used?

ANTONIO
Robotics Engineer

STEM Career Connection
Materials Engineer

June 12

A local cargo company asked me to analyze the stability of their shipping barges. My goal is to help the cargo company move their materials by using less fuel. From what I can tell, the barges are overloaded and are displacing a lot of water, making it more difficult for them to move smoothly.

June 27

I suggested they load less cargo onto each barge or invest in barges made out of lighter weight material. That way, less fuel will be needed to move the ships at the speed required to get them to their destination on time.

Draw how you think the cargo should be loaded on the barge.

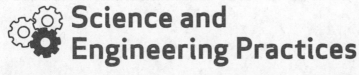

Science and Engineering Practices

I will plan and carry out investigations.

I will develop and use models.

Name _____ Date _____

Matter's Structure

Is It Matter?

Four friends were talking about matter. They each had different ideas about the kinds of things that are matter. This is what they said:

Abe: *I think something needs to be solid to be matter.*

Kayla: *I think matter can be a solid or a liquid.*

Curtis: *I think matter can be a solid, liquid, or gas.*

Lori: *I think matter can be a solid, liquid, or gas, but it doesn't include living things.*

Who do you agree with most? _____

Explain why you agree.

_____

_____

_____

_____

Science in Our World

▶ Watch the video of the iceberg. What is it made of?
What questions do you have about this mass of ice in the ocean?

_____

_____

_____

_____

Read about a physics teacher and answer the questions on the next page.

> Physics teachers can use models to help students understand the structure of different types of matter.

STEM Career Connection
Physics Teacher

I'm a physics teacher at Roosevelt High School, and I love what I do! There is much to explore in the world of physics, and although we have to take notes and do complex math, my students have fun. They work hard and even use computer models to show what they have learned.

My favorite days are when we do investigations. Sometimes I have all the materials ready to go and the students make predictions and follow instructions to complete the experiment. But sometimes I let them design their own experiments based on questions they want to answer about the physical world. These days are great. Sometimes they ask questions and find answers that I never thought of. And that is what science is all about—finding ways to discover new things!

HANNAH
Welder

1. What does the physics teacher help students do?

_____

_____

_____

2. What is a tool physics students might use? How would they use it?

_____

_____

_____

_____

? Essential Question

How are the particles in matter organized?

_____

_____

_____

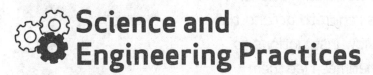

Science and Engineering Practices

I will develop and use models.

Like a physics teacher, you will develop a model to show the structure of matter.

Science in Our World

▶ Watch the video of the iceberg. What is it made of?
What questions do you have about this mass of ice in the ocean?

_____

_____

_____

_____

Read about a physics teacher and answer the questions on the next page.

> Physics teachers can use models to help students understand the structure of different types of matter.

STEM Career Connection
Physics Teacher

I'm a physics teacher at Roosevelt High School, and I love what I do! There is much to explore in the world of physics, and although we have to take notes and do complex math, my students have fun. They work hard and even use computer models to show what they have learned.

My favorite days are when we do investigations. Sometimes I have all the materials ready to go and the students make predictions and follow instructions to complete the experiment. But sometimes I let them design their own experiments based on questions they want to answer about the physical world. These days are great. Sometimes they ask questions and find answers that I never thought of. And that is what science is all about—finding ways to discover new things!

HANNAH
Welder

1. What does the physics teacher help students do?

_____

_____

_____

_____

2. What is a tool physics students might use? How would they use it?

_____

_____

_____

_____

? Essential Question

How are the particles in matter organized?

_____

_____

_____

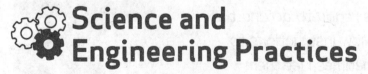

Science and Engineering Practices

I will develop and use models.

Like a physics teacher, you will develop a model to show the structure of matter.

Measuring Matter

▶ Watch *Measuring Matter* on different ways to measure matter.

📖 Then, read pages 260–261 in the *Science Handbook.* Answer the following question after you have finished reading.

4. How can making observations and measurements of matter help us determine its structure?

_____

_____

_____

Particles in Matter

🔊 Explore the Digital Interactive *Particles in Matter* on how the properties of matter are related to the structure of its particles.

📖 Then, read pages 256–257 in the *Science Handbook.* Answer the following questions after you have finished reading.

⚙ Crosscutting Concepts
Scale, Proportion, and Quantity

5. Compare the particles in the equal amounts of a solid, a liquid, and a gas.

_____

_____

_____

_____

_____

6. How does the organization of particles in matter relate to the matter's state?

_____

_____

_____

_____

_____

Science and Engineering Practices

Think about how you have used models to understand the structure of the different states of matter. Tell how you can use a model of matter's structure by completing the "I can . . ." statement below.

I can _____

_____

_____

Use examples from the lesson to explain what you can do!

Object	Prediction: Sink or Float?	Observation: Sink or Float?	Mass (g)	Volume (mL)	Density (g/mL)
Golf Ball					
Table Tennis Ball					
Marble					

Communicate Information

4. How does calculating the density of each object help explain whether it sinks or floats in water?

_____

_____

_____

_____

5. **Make an Argument** Choose one of the objects from the activity. Use evidence from the investigation to explain why it either sinks or floats when placed in water.

_____

_____

_____

_____

_____

Performance Task
Modeling Matter

Think like a physics teacher and use what you have learned about the structure of matter to develop models to show how particles are arranged in the three main states of matter.

Define a Problem How can you use the materials provided to model the three states of matter?

_____

_____

_____

_____

_____

_____

Materials
☐ 3 paper plates
☐ marker
☐ 100 small, same-size objects (buttons, beads, beans)
☐ glue

Make a Model

1. Label each of the three paper plates as Solid, Liquid, or Gas. The plates represent the same amount of space that the particles will take up.

2. Think about the organization of particles in a solid, liquid, and gas. Divide the small objects into three groups to represent the particles in each state of matter.

3. Place the objects that represent the solid's particles on the plate labeled Solid. Glue the objects on the plate to represent the arrangement of the particles in a solid.

4. Repeat step 3 for the plate labeled Liquid and Gas.

Think like a physics teacher and develop models to represent the structures of the three states of matter.

5 **Record Data** Create a sketch of each model. Explain each sketch.

Communicate Information

1. **Construct an Explanation** What does the paper plate represent in the model of each of the three states of matter?

_____

_____

_____

2. **Construct an Explanation** Which of your models is the most dense? How do you know?

_____

_____

_____

_____

3. Describe how you could use your models to teach younger students about the structure of matter.

_____

_____

_____

_____

_____

_____

_____

_____

_____

? Essential Question
How are the particles in matter organized?

▶ Think about the video of the iceberg at the beginning
of the lesson. Explain how the structure of matter in the iceberg
compares to the structure of matter in the ocean and why
the iceberg is able to float.

_____

_____

_____

_____

_____

_____

_____

Science and Engineering Practices

Now that you're done
with the lesson, share
what you did!

Review the "I can . . ." statement you wrote earlier in the
lesson. Explain what you have accomplished in this lesson by
completing the "I did . . ." statement.

I did _____

_____

_____

_____

_____

_____

Matter's Properties

Particles of Matter

Four friends were talking about the particles that make up matter and give matter its properties. They each had different ideas. This is what they said:

Joyce: I think you can't see the particles that make up solids, liquids, and gases. They are too small to see.

Harold: I think you can see the particles that make up solids, liquids, and gases.

Royce: I think you can see the particles that make up solids and liquids, but you can't see the particles that make up gases.

Benito: I think you can see the particles that make up solids, but you can't see the particles that make up liquids and gases.

Who do you agree with most? _____

Explain why you agree.

_____

_____

_____

_____

Science in Our World

Look at the photo of the artist blowing glass. How does this occur?
What questions do you have about this process?

_____

_____

_____

_____

Read about a materials scientist and answer the questions
on the next page.

> Materials scientists investigate the properties of matter to determine how it can be used.

STEM Career Connection
Materials Scientist

People have been working with metal for
thousands of years. Even parts of human history
have been named after the metals that were commonly
used. Have you heard of the Iron Age or the Bronze
Age? Today, materials scientists are using metals in
ways that no one could have pictured before. Can you
believe they can shape a copper wire thousands of
times thinner than a strand of your hair?

Materials scientists try to combine elements and
compounds into new materials with useful properties.
In the past, people used and changed materials by
trial and error. Modern materials scientists combine
and change materials based on an understanding of
their properties and how matter is put together. They
often work with materials on a very small scale.

HANNAH
Welder

1. How does a materials scientist investigate matter?

_____

_____

_____

_____

2. How can materials scientists combine different materials more efficiently than they did in the past to make them more useful?

_____

_____

_____

_____

? Essential Question

How do the particles in matter affect its properties?

_____

_____

_____

⚙ Science and Engineering Practices

I will plan and carry out an investigation.

Like a materials scientist, you will carry out investigations to explore the properties of matter.

Inquiry Activity
What Is Inside Matter?

How can you determine the properties of different types of matter in sealed boxes without being able to see them?

Make a Prediction What three tests could you do to determine what is in the boxes without opening them?

_____

_____

_____

_____

Materials

☐ four sealed boxes

☐ pan balance

☐ other testing items as needed

Carry Out an Investigation

1 **Record Data** Your teacher will provide four sealed boxes. Identify each test you will perform in the data table under Test 1, Test 2, and Test 3.

2 Perform the tests, one at a time, on each of the four boxes. Record the results.

3 **Make a Prediction** Record your prediction of what you think is in each box.

	Test 1	Test 2	Test 3	Prediction
Box 1				
Box 2				
Box 3				
Box 4				

Communicate Information

1. What test was the most useful for making a prediction of what object was in each box?

_____

_____

_____

2. Open each of the boxes. Were your predictions correct? Explain.

_____

_____

_____

3. **Construct an Explanation** Is it possible to have accurate observations but still predict incorrectly? Explain.

_____

_____

_____

_____

_____

4. How else could you determine the type of matter in each box without seeing it?

_____

_____

_____

Obtain and Communicate Information

abc Vocabulary

> **Use these words when explaining the properties of matter.**
>
> element   compound
>
> atom   molecule

Observing and Measuring Matter, Elements, and Atoms

📖 Read pages 262–265 in the *Science Handbook.* Answer
the following questions after you have finished reading.

1. Which is smaller, a molecule or an atom? Explain your answer.

_____

_____

_____

2. How are elements and compounds related?

_____

_____

_____

3. How are compounds formed? Give an example of a compound.

_____

_____

_____

_____

FOLDABLES®

Cut out the Notebook Foldables tabs given to you by your teacher. Glue the anchor tabs as shown below. Use what you have learned about the properties of matter.

Glue anchor tab here

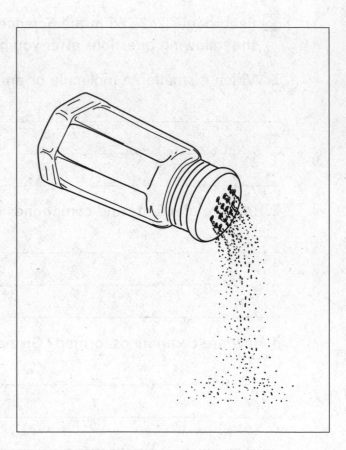

Elements, Atoms, and Molecules

👁 Read *Elements, Atoms, and Molecules* on the building blocks of matter. Answer the following questions after you have finished reading.

⚙ Crosscutting Concepts
Scale, Proportion, and Quantity

4. What is the smallest unit of an element that has the properties of that element?

_____

_____

5. What forms when two or more atoms join together?

_____

_____

6. Describe how elements, atoms, and molecules are related.

_____

_____

_____

_____

Chemical Properties

📖 Read pages 258–259 in the *Science Handbook.* Answer the following questions after you have finished reading.

7. What is the difference between a physical property and a chemical property?

_____

_____

_____

_____

8. How could a materials scientist test for chemical properties?

_____

_____

_____

_____

Properties of Elements

Explore the Digital Interactive *Properties of Elements* on the properties of some elements in the periodic table. Answer the question after you have finished.

9. Is aluminum's ability to be pressed into sheets a physical property or chemical property?

_____

_____

Science and Engineering Practices

Think about how you have investigated the properties of matter. Tell how you can carry out an investigation to identify the properties of matter by completing the "I can . . ." statement below.

Use examples from the lesson to explain what you can do!

I can _____

_____

_____

_____

Research, Investigate, and Communicate

Using Elements

Explore the Digital Interactive *Using Elements* on how certain elements are used based on their properties. Answer the questions after you have finished.

1. Why is helium used in balloons?

_____

_____

_____

_____

2. Why is it important to know the properties of matter when deciding how to use it?

_____

_____

_____

_____

_____

Model a Compound

Research Research an everyday compound, such as water or salt. Explain how the properties of the compound are different from the properties of the elements that make up the compound. For example, hydrogen and oxygen are gases at room temperature; however, they combine to form H_2O, or water, which is liquid at room temperature. Record your notes below.

_____

_____

_____

_____

_____

Make a Model Draw a labeled diagram of your compound here.

With materials provided by your teacher, make a 3-dimensional model of your molecule.

⚙ Performance Task
Testing Matter's Properties

You will use skills, like a materials scientist, to measure and record data about different types of matter and how they interact.

Make a Prediction Do you think each of the solid materials will interact the same way with each of the liquids? Explain.

_____

_____

_____

_____

Carry Out an Investigation

BE CAREFUL Wear safety goggles and use caution to avoid spills. Do not taste any of the materials.

1. Use the balance to measure 5 grams (g) of baking soda on a square of wax paper. Carefully pour the baking soda into the first section of the egg carton. Label the section 'baking soda.'

2. Repeat step 1 two more times for baking soda so that you have three sections each with 5 g of baking soda in them.

3. Repeat steps 1 and 2 for the baking powder and for the salt. Label the sections. You should have 3 sections with 5 g each of baking soda, 3 with 5 g each of baking powder, and 3 with 5 g each of salt.

4. **Record Data** Use the graduated cylinder to measure 20 mL of water. Pour it into one of the sections with baking soda. Record your observations in the table on the next page.

5. Pour 20 mL of vinegar into one of the remaining two baking soda sections. Record your observations.

6. Pour 20 mL of iodine solution into the remaining baking soda section. Record your observations.

Materials
☐ safety goggles
☐ 15 g baking soda
☐ wax paper
☐ pan balance
☐ empty egg carton
☐ marker
☐ 15 g baking powder
☐ 15 g salt
☐ 60 mL water
☐ graduated cylinder
☐ 60 mL vinegar
☐ 60 mL iodine solution

7 Repeat steps 4-6 for the baking powder and salt sections.

Observations	Water	Vinegar	Iodine Solution
Baking soda			
Baking powder			
Salt			

Communicate Information

1. **Analyze Data** How were the interactions alike? How were they different?

_____

_____

_____

_____

2. **Construct an Explanation** What properties were you able to observe about these materials during the investigation?

_____

_____

_____

_____

_____

_____

_____

❓ Essential Question
How do the particles in matter affect its properties?

Think about the photo of an artist blowing glass. If you were a materials scientist, how would you use the properties of matter to explain how this can occur?

_____

_____

_____

_____

_____

_____

_____

⚙ Science and Engineering Practices

Now that you're done with the lesson, share what you did!

Review the "I can . . ." statement you wrote earlier in the lesson. Explain what you have accomplished in this lesson by completing the "I did . . ." statement.

I did _____

_____

_____

_____

_____

Name _____ Date _____

Metals and Nonmetals

 **PAGE KEELEY SCIENCE PROBES**

Properties of Metals

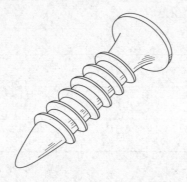

There are many different types of metals and nonmetals. Both can be described by their properties. Put an X in any of the boxes that describe a property that can be used to tell a metal from a nonmetal.

shiny	mass	solid
weight	good conductor of electricity	some are magnetic
hard	length	good conductor of heat
soft	can be hammered or pressed without breaking	sharp

Explain your choices.

_____

_____

_____

_____

Science in Our World

Look at the photo of the magnet and iron filings. Why do you think the filings are arranged this way? What questions do you have about these objects?

_____

_____

_____

_____

Read about a mechanical engineer and answer the questions on the next page.

> Mechanical engineers need to consider the properties of metals and nonmetals when they design and build new machines.

STEM Career Connection
Mechanical Engineer

Have you ever been on a roller coaster? Chances are good that a mechanical engineer designed it. I combine physics, engineering, and materials science to design exciting and safe roller coasters. Using what I know about the properties of metals and nonmetals allows me to make sure I choose the right materials that will be used to build the coaster.

That fast start to the ride is accomplished by using very powerful electromagnets that repel each other when turned on. No more gradual climbs up a big hill. Just buckle in and . . . whoosh! Magnets are also used to quickly but gently slow you down at the end of the ride.

HANNAH
Welder

1. What does a mechanical engineer need to know in order to design a roller coaster?

_____

_____

_____

_____

_____

2. How are magnets used on roller coasters?

_____

_____

_____

_____

_____

? Essential Question
What are some properties of metals and nonmetals?

_____

_____

_____

_____

⚙ Science and Engineering Practices

I will develop and use a model.

> Like a mechanical engineer, you will develop a model of a solution that uses the properties of metals and nonmetals.

Inquiry Activity
Metal or Not?

How can you tell whether a material has the properties of a metal or nonmetal?

Make a Prediction Which properties being tested in the table do you think are those of metals?

_____

_____

_____

_____

Carry Out an Investigation

BE CAREFUL Wear safety goggles. The lightbulb will get hot! Do not touch the lightbulb or the part of each object that directly touches the lightbulb. Be careful with any sharp or pointy edges on the materials.

1 **Record Data** Hold each object (one at a time) against the lightbulb for one minute. Did the end you were holding get warm? Record the results in the table on the next page.

2 Pick up each object and gently try to bend it. Did it bend? Record the results.

3 Touch the magnet to each of the objects. Did it stick? Record the results.

4 Hold each object under the light. Is it shiny? Record the results.

5 Discuss the results with your classmates. Try to decide whether each object is a metal based on the results of the tests. Record your decisions.

Materials

- [ ] safety goggles
- [ ] acrylic rod
- [ ] paper clip
- [ ] rubber eraser
- [ ] paper
- [ ] aluminum foil
- [ ] copper wire
- [ ] wooden toothpicks
- [ ] magnet
- [ ] lamp with incandescent lightbulb

Object	Did it get warm?	Could you bend it?	Did the magnet stick to it?	Is it shiny?	Metal or not?
Acrylic Rod					
Paper Clip					
Rubber Eraser					
Paper					
Aluminum Foil					
Copper Wire					
Wooden Toothpick					

Communicate Information

1. Were some objects harder or easier to conclude as metals? Explain.

_____

_____

2. Which property was the easiest to determine?

_____

_____

3. Which property was the most difficult to determine? Why?

_____

_____

4. Were your predictions correct? Explain why or why not.

_____

_____

_____

Obtain and Communicate Information

Vocabulary

Use these words when explaining metals and nonmetals.

metal	nonmetal	conductivity
ductility	malleability	magnetism
corrosion	alloy	

Organization of the Periodic Table

Explore the Digital Interactive *Organization of the Periodic Table* on how metals and nonmetals are arranged in the periodic table. Answer the question after you have finished.

1. How is the periodic table organized in terms of metals and nonmetals?

_____

_____

_____

Properties of Metals

Explore the Digital Interactive *Properties of Metals* on the different properties of metals. Answer the questions after you have finished.

2. Why are metals conductors?

_____

_____

_____

_____

_____

3. What does it mean for metals to be malleable?

_____

_____

_____

_____

Physical Properties

Read pages 254–255 in the *Science Handbook.* Answer the following question after you have finished reading.

4. Which properties of matter are shown by metals?

_____

_____

_____

_____

Metals, Nonmetals, and Metalloids

Read *Metals, Nonmetals, and Metalloids* about different types of matter. Answer the following question after you have finished reading.

5. Why are metalloids important to use in modern technology?

_____

_____

_____

_____

_____

FOLDABLES®

Cut out the Notebook Foldables tabs given to you by your teacher. Glue the anchor tabs as shown below. Use what you have learned to explain the properties of metals and nonmetals.

Glue anchor tab here

Glue anchor tab here

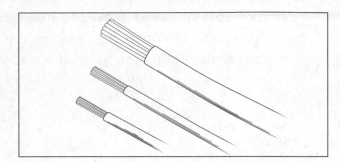

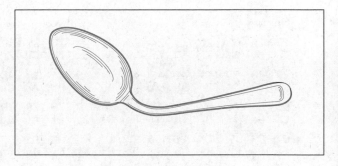

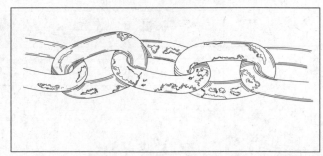

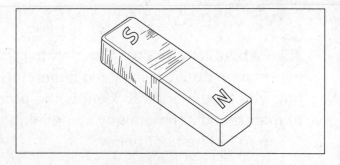

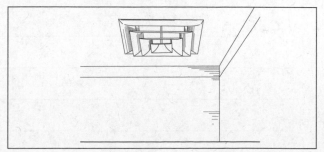

Properties of Metals, Nonmetals, and Metalloids

After completing the *Properties of Metals* interactive and reading *Metals, Nonmetals, and Metalloids*, color code the periodic table below to show the three different groups. Use colored pencils to lightly shade all of the metals yellow, all of the nonmetals green, and all of the metalloids blue.

1																		18
1 H	2												13	14	15	16	17	2 He
3 Li	4 Be												5 B	6 C	7 N	8 O	9 F	10 Ne
11 Na	12 Mg	3	4	5	6	7	8	9	10	11	12		13 Al	14 Si	15 P	16 S	17 Cl	18 Ar
19 K	20 Ca	21 Sc	21 Ti	23 V	24 Cr	25 Mn	26 Fe	27 Co	28 Ni	29 Cu	30 Zn		31 Ga	32 Ge	33 As	34 Se	35 Br	36 Kr
37 Rb	38 Sr	39 Y	40 Zr	41 Nb	42 Mo	43 Tc	44 Ru	45 Rh	46 Pd	47 Ag	48 Cd		49 In	50 Sn	51 Sb	52 Te	53 I	54 Xe
55 Cs	56 Ba	57- 71	72 Hf	73 Ta	74 W	75 Re	76 Os	77 Ir	78 Pt	79 Au	80 Hg		81 Tl	82 Pb	83 Bi	84 Po	85 At	86 Rn
87 Fr	88 Ra	89- 103	104 Rf	105 Db	106 Sg	107 Bh	108 Hs	109 Mt	110 Ds	111 Rg	112 Cn		113 Uut	114 Fl	115 Uup	116 Lv	117 Uus	118 Uuo

58 Ce	59 Pr	60 Nd	61 Pm	62 Sm	63 Eu	64 Gd	65 Tb	66 Dy	67 Ho	68 Er	69 Tm	70 Yb	71 Lu
90 Th	91 Pa	92 U	93 Np	94 Pu	95 Am	96 Cm	97 Bk	98 Cf	99 Es	100 Fm	101 Md	102 No	103 Lr

Science and Engineering Practices

Think about how you used models to gather information about metals and nonmetals. Tell how you used models to identify the properties of metals and nonmetals by completing the "I can . . ." statement below.

Use examples from the lesson to explain what you can do!

I can _____

_____

_____

◨ᵕ Research, Investigate, and Communicate

Researching Metals

Most metals that conduct electricity well also conduct heat well, and those that do not conduct electricity well do not conduct heat well.

📖 **Research** Use your *Science Handbook* and other research materials to determine which types of metals conduct electricity and heat better than others.

⚙ Crosscutting Concepts
Scale, Proportion, and Quantity

1. Choose three metals from the periodic table and research their conductivity. List them in order based on how conductive they are, and describe common uses for each of the metals.

_____

_____

_____

_____

_____

_____

_____

2. **Construct an Explanation** Choose one of the metals you researched. Based on its properties, what new uses you can think of for this metal? What uses of this metal should be avoided?

_____

_____

_____

Performance Task
Using Metals and Nonmetals

You will use what you have learned about the properties of metals and nonmetals to think like a mechanical engineer.

Define a Problem How can you use both metal and nonmetal materials to design a solution that makes a task easier?

_____

_____

_____

Design a Solution

Use the space below to plan your solution and draw a model of what it would look like. If you have time and materials, make a 3-dimensional version of your model.

Think like a mechanical engineer and model a solution that uses both metals and nonmetals.

Communicate Information

1. What task does your solution make easier? How?

_____

_____

_____

_____

_____

2. What materials will you use for your solution?

_____

_____

_____

_____

_____

3. What properties of metals and nonmetals did you consider
when making your design?

_____

_____

_____

_____

_____

_____

? Essential Question
What are some properties of metals and nonmetals?

Think about the photo of the iron filings and magnet. You now know that some metals are magnetic, which is why the filings are arranged this way. What are some other properties of metals?

_____

_____

_____

_____

_____

⚙ Science and Engineering Practices

Review the "I can . . ." statement you wrote earlier in the lesson. Explain what you have accomplished in this lesson by completing the "I did . . ." statement.

Now that you're done with the lesson, share what you did!

I did _____

_____

_____

_____

_____

_____

Structure and Properties of Matter

⚙ Performance Project
Build a Better Boat

Define a Problem Use what you have learned about the structure and properties of matter to build a boat. Make sure that the structure of the boat and the building materials you choose can hold the weight of the cargo and transport the goods to their destination. Write this problem in your own words.

_____

_____

_____

Design a Solution

BE CAREFUL Use caution to not spill any water.

① **Make a Prediction** Which material will make the best boat? Explain.

_____

_____

② Draw a sketch of the boat design you intend to use to solve the problem.

Structure and Properties of Matter

3 Choose a material and build your boat.

4 **Test Your Solution** Place the boat in a large pan of water. Place paper clips, one at a time, into the boat and record what happens.

5 **Record Data** Record the number of paper clips the boat could hold on the chart. Record other observations.

6 Repeat steps 3–5 with each type of material.

Materials

- ☐ aluminum foil
- ☐ construction paper
- ☐ wax paper
- ☐ clay
- ☐ paper clips
- ☐ large pan of water

Material	Number of Paper Clips	Observation
Aluminum Foil		
Construction Paper		
Wax Paper		
Clay		

How can I use what I know about different types of matter to show how they can be used to build a boat?

Communicate Information

1. **Analyze Data** Which boat held the most paper clips? What does this tell you about the properties of the material you used to build that boat?

_____

_____

_____

_____

What patterns did you notice between the structure and property of the material used to build each boat and the amount of cargo it can hold?

2. **Improve Your Design** How would you change your boat design so that the boat could hold more mass? Draw your changes below.

 ## Explore More in Our World

Did you learn the answers to all of your questions from the beginning of the module? If not, how could you design an experiment or conduct research to help answer them?

Physical and Chemical Changes

 ## Science in Our World

Matter comes in different forms, and sometimes we change its form
in order to create a solution to a problem. Look at the photo of
molten metal being poured. What questions do you have about
this process?

_____

_____

_____

_____

_____

abc Key Vocabulary

Look and listen for these words as you learn about physical and chemical changes.		
boiling point	chemical change	chemical property
conservation of mass	freezing point	melting point
mixture	physical change	product
reactant	solubility	solution

How can I investigate the differences between matter dissolving and matter melting?

HANNAH
Welder

STEM Career Connection

Metallurgist

June 1

We've heated the iron ore up to 1300°C. That's hot, but not quite hot enough. We'll need to get it up to 1538°C before it melts. People have been working with iron for thousands of years, but only recently have we understood how the arrangement of atoms inside the iron changes as it is heated and cooled.

The new arrangement of atoms will make it a much stronger iron than when we started if we let it cool. We want to bring it to its melting point. Steel is a mixture of iron and other elements. Basic steel is 99% iron and 1% carbon, and it is much stronger than 100% iron. Other metals can be mixed with iron to form different kinds of steel, each with different properties.

How can iron be made into a stronger metal?

_____

_____

_____

Science and Engineering Practices

I will plan and carry out investigations.

I will use mathematics and computational thinking.

Physical Changes

PAGE KEELEY
**SCIENCE
PROBES**

What Happened to the Mass?

John's mother put chocolate chip cookies in his backpack.
The cookies were in a resealable bag on the bottom of John's
backpack. John put his books in his backpack and went to school.
At snack time, he took out the bag and noticed that the cookies
were smashed into crumbs. He wondered whether the mass of the
cookies changed. What do you think? Circle the answer that
best matches your thinking.

A. *The mass of the whole cookies is the same as the mass of
all the crumbs.*

B. *The mass of the whole cookies is different from the mass of
all the crumbs.*

Explain your thinking. What rule or reasoning helped you decide what happened to
the mass?

_____

_____

_____

_____

_____

Science in Our World

Look at the frozen waterfall. What caused it to freeze? What questions do you have about how this waterfall compares to a flowing waterfall?

_____

_____

_____

Read about a civil engineer and answer the questions on the next page.

A civil engineer needs to consider how water freezes and melts when designing the structure.

STEM Career Connection
Civil Engineer

As a civil engineer, I have a lot to consider when designing a dam in a river or stream. These structures are important for controlling the flow of water in areas where people live. Dams help prevent flooding, irrigate crops, and provide water for humans to use. The flow of water through a dam is also used in generating hydropower.

The physical changes that water goes through, such as freezing when the temperature is very low, can cause wear and tear on the dam's structure. We have to monitor the dam regularly and repair damaged areas to make sure nothing is stopping the flow of water.

MALIK
Photonics Engineer

1. Why are dams important structures?

_____

_____

_____

_____

2. What do you wonder about the role of a civil engineer?

_____

_____

_____

_____

❓ Essential Question
What happens to the amount of matter when it changes state?

_____

_____

_____

_____

⚙️ Science and Engineering Practices

I will carry out an investigation.

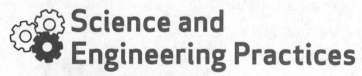

Like a civil engineer, you will investigate whether the mass of matter is affected by its state.

Inquiry Activity
Frozen or Unfrozen

Does the mass of matter change when it changes state?

Make a Prediction Do you think the mass and volume of water will change when it freezes? Why or why not?

_____

_____

_____

_____

Carry Out an Investigation

1. **Record Data** Measure and record the mass of the bottle of liquid water in the table.

2. Repeat step 1 for the bottle with frozen water.

3. Examine the two bottles side by side. Record any differences in the apparent volume by looking at height of the surface of the water in each bottle and the shape of the bottle. Record this in the table.

4. Draw a line at the surface of the water in the liquid and frozen water bottles. Revisit the bottles after the frozen one has melted to see whether there are any changes.

	Mass	Apparent Volume
Liquid Water		
Frozen Water		

Communicate Information

1. Did freezing the water change its mass? Explain.

_____

_____

2. Did freezing the water change its volume? Explain.

_____

_____

_____

3. What does this activity tell you about how matter is conserved in a physical change?

_____

_____

_____

4. How could you make the frozen water bottle melt faster? Would this change the mass of the water in the bottle?

_____

_____

_____

_____

FOLDABLES®

Cut out the Notebook Foldables tabs given to you by your teacher. Glue the anchor tabs as shown below. Describe the three states of matter in your own words and draw examples of each.

Glue anchor tab here

  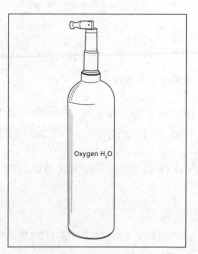

Obtain and Communicate Information

Vocabulary

Use these words when explaining physical changes in matter.

boiling point   freezing point   physical change

conservation of   melting point
  mass

Physical Properties

📖 Read pages 254–257 in the *Science Handbook* to review physical properties of matter. Answer the following question after you have finished reading.

1. Give an example of a physical change that would change a physical property of matter.

_____

_____

_____

_____

Changes in Matter

📖 Read pages 266–267 in the *Science Handbook.* Answer the following questions after you have finished reading.

2. What is needed for matter to melt?

_____

3. Is matter changing state an example of a physical or a chemical change? Why?

_____

_____

_____

4. How is heat energy related to an object changing state?

_____

_____

_____

Mass and Volume

👁 Read *Mass and Volume* on whether volume in matter changes during a physical change. Answer the questions after you have finished reading.

5. Does mass change as the volume of matter changes? Explain.

_____

_____

_____

_____

6. What would happen to the water vapor from the pot of boiling water as it begins to lose heat energy?

_____

_____

_____

Particles in Matter

▦ Investigate the particles in different types of matter by conducting the simulation. Answer the following questions when you have finished.

⚙ Crosscutting Concepts
Cause and Effect

7. What happens to the particles in iodine when you increase the temperature to 200 degrees Celsius?

_____

_____

_____

8. What happens to the particles in olive oil when you increase
the temperature to 200 degrees Celsius? What does this tell you?

_____

_____

_____

_____

_____

_____

9. What happens to the mass of the materials regardless of whether
heat is added or removed?

_____

_____

⚙ Science and Engineering Practices

Think about how you have investigated if the amount
of matter changes in a physical change. Tell how you
can carry out an investigation by completing the
"I can . . ." statement below.

Use examples from the lesson to explain what you can do!

I can _____

_____

_____

_____

🔍 Research, Investigate, and Communicate

Temperature Points

🔁 Explore the Digital Interactive *Temperature Points* on the melting, boiling, and freezing points of different types of matter. Answer the questions after you have finished.

1. Why is water a unique example of matter on Earth?

   _____

   _____

   _____

2. What is the boiling point of water?

   _____

   _____

3. What is the freezing point of water?

   _____

   _____

4. Do you think the amount of water on Earth ever changes? Explain.

   _____

   _____

   _____

Performance Task
Build a Dam

You will use what you know about the physical properties of water to design a dam with a variety of materials. Then, you will investigate whether the amount of water changes as it flows through the dam.

Make a Prediction Will the flow of water through the dam result in a change in the amount of water? Explain.

_____

_____

_____

_____

Materials
☐ safety goggles
☐ scissors
☐ cardboard
☐ plastic trough or planter
☐ straws
☐ plastic tubing
☐ plastic wrap
☐ tape
☐ modeling clay
☐ large beaker
☐ water

Carry Out an Investigation

BE CAREFUL Wear safety goggles. Use caution when using scissors.

1. Cut a piece of cardboard so that it fits snugly across the width of the plastic container.

2. Decide how you will allow water to flow through the dam. You can use the straws, tubing, and other materials. Draw your design on the next page.

3. Build the dam using your design. Use the plastic wrap, tape, and modeling clay to secure the cardboard across the width of the container.

4. Fill a large beaker with 500 mL of water.

5. Carefully pour the water into one side of the dam. Fill up that side of the container so the water starts to flow through the dam. Keep track of the amount of water you pour into your model. Note how many times you fill up the beaker to fill one side.

6. Observe how the water flows through the dam.

7. Carefully scoop the water from the other side of the dam. Record how much water you take out.

Think like a civil engineer and explore if water changes as it moves through the dam.

Draw your design to build your dam in the box below.
Label the materials you will use in your design.

Communicate Information

1. **Construct an Explanation** Did the amount of water change as it went through the dam? Why or why not?

_____

_____

_____

_____

2. **Construct an Explanation** Did the water undergo a physical change? Why or why not?

_____

_____

_____

_____

3. How would you need to change your model to get frozen water to flow through the dam?

_____

_____

_____

_____

_____

_____

_____

? Essential Question

What happens to the amount of matter when it changes state?

Think about the photo of the frozen waterfall from the beginning of the lesson. Explain why the amount of water in the waterfall does not change if it freezes.

_____

_____

_____

_____

Science and Engineering Practices

Review the "I can . . ." statement you wrote earlier in the lesson. Explain what you have accomplished in this lesson by completing the "I did . . ." statement.

Now that you're done with the lesson, share what you did!

I did _____

_____

_____

_____

_____

_____

_____

Mixtures and Solutions

PAGE KEELEY
SCIENCE
PROBES

Salt and Water

A spoonful of salt has a mass of 10 grams. A cup of water has
a mass of 300 grams. What do you predict will be the total
mass of the saltwater when the salt is dissolved in the water?
Circle the answer that best matches your thinking.

A. *more than 300 grams*

B. *less than 300 grams*

C. *300 grams*

Explain your thinking. What rule or reasoning did you use to make your prediction?

_____

_____

_____

_____

_____

Science in Our World

Look at the photo of the solid dissolving in the liquid. Are all solids able to dissolve? What questions do you have about this mixture?

_____

_____

_____

_____

Read about a pharmacist and answer the questions on the next page.

> A pharmacist needs to be precise when mixing matter to prepare medicines.

STEM Career Connection
Pharmacist

As a pharmacist, I am an expert on how different medications affect the human body. But it's more than just putting pills in a bottle! Many medicines are dangerous if the dosage is too strong and not effective if the dosage is too weak. I have to make sure the dosage is correct for the patient's body weight.

A way to do that is to dissolve the correct amount of medicine in a solution. One type of a solution is where a solid is dissolved in a liquid. I also have to know what other medications a patient might be taking so that I can be sure the medicines do not react with one another.

MALIK
Photonics Engineer

1. Why do pharmacists need to use measurements?

_____

_____

_____

2. What else do you wonder about the role of a pharmacist?

_____

_____

_____

? Essential Question

What happens to the mass when different types of matter are mixed?

_____

_____

_____

⚙ Science and Engineering Practices

I will use mathematics and computational thinking.

> Like a pharmacist, you will use measurements to describe what happens when matter is mixed.

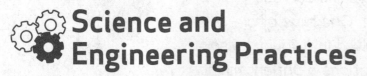

Inquiry Activity
Solubility Solutions

Does temperature affect how easily something can be mixed?

Make a Prediction Will the solids dissolve faster in warm water or cold water? Explain your reasoning.

_____

_____

_____

_____

Materials

☐ safty goggles

☐ 2 graduated cylinders

☐ 400 mL warm water

☐ 400 mL cold water

☐ pan balance

☐ wax paper

☐ 20 g sugar

☐ stopwatch

☐ stirrers

☐ 20 g salt

Carry Out an Investigation

BE CAREFUL Wear safety goggles. Use caution to avoid spills.

1. **Record Data** Pour 200 milliliters (mL) of warm water into a graduated cylinder. Measure the mass of the graduated cylinder with water in it. Record the mass and volume in the table on the next page.

2. Repeat step 1 with cold water in a second graduated cylinder.

3. Pour 20 grams (g) of sugar into the warm water.

4. Start the stopwatch and begin stirring the sugar into the warm water. Stop the timer when the sugar dissolves. Measure the volume of the mixture. Record the time and volume.

5. Repeat steps 3 and 4 with the cold water.

6. Measure the mass of each graduated cylinder again. Record the data in the table.

7. **Record Data** Repeat steps 1-6, using salt instead of sugar. Record the data in the table.

Name _____ Date _____

	Mass	Volume	Time to Mix
Cold Water			
Sugar			
Cold Water + Sugar			
Warm Water			
Warm Water + Sugar			
Salt			
Cold Water + Salt			
Warm Water + Salt			

Communicate Information

1. In which temperature of water did the sugar dissolve faster?

_____

2. Was there a difference between the mass and volume of the cold water and the warm water? Explain.

_____

_____

3. In which temperature of water did the salt dissolve faster?

_____

_____

4. Was your prediction correct? Why or why not?

_____

_____

_____

Obtain and Communicate Information

abc Vocabulary

> **Use these words when explaining mixtures and solutions.**
>
> mixture   solution   solubility
>
> colloid   distillation

Types of Mixtures

📖 Read pages 268–271 in the *Science Handbook*. Answer the following questions after you have finished reading.

1. Explain the difference between heterogeneous and homogenous mixtures.

_____

_____

_____

_____

_____

2. What is the difference between a suspension and a colloid?

_____

_____

_____

_____

3. How is the solubility of a material determined?

_____

_____

_____

FOLDABLES®

Cut out the Notebook Foldables tabs given to you by your teacher. Glue the anchor tabs as shown below. Use what you have learned to describe different mixtures.

Glue anchor tab here

Mayonnaise

Mixtures in Action

Investigate what happens when different types of matter are mixed by conducting the simulation. Answer the questions after you have finished.

Crosscutting Concepts
Cause and Effect

4. What happens to the mass of two materials when they are mixed?

_____

_____

_____

_____

5. What type of mixture do you think is created when the water and gelatin are mixed together?

_____

_____

6. What might have caused the salt to not completely dissolve in the water?

_____

_____

_____

_____

Science and Engineering Practices

Use examples from the lesson to explain what you can do!

Think about how you have used mathematics and computational thinking to describe conservation of mass in mixtures by completing the "I can . . ." statement below.

I can _____

_____

_____

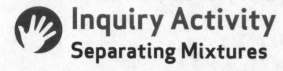

Research, Investigate, and Communicate

Inquiry Activity
Separating Mixtures

You will investigate if and how a mixture can be separated.

Write a Hypothesis Will different methods of separating mixtures work equally well on the same mixture? Write your answer in the form of an "If...then..." statement.

_____

_____

_____

_____

<table>
<tr><td>**Materials**</td></tr>
<tr><td>☐ safety goggles</td></tr>
<tr><td>☐ large bowl</td></tr>
<tr><td>☐ 1 cup of sand</td></tr>
<tr><td>☐ 1/2 cup of salt</td></tr>
<tr><td>☐ 200 mL of water</td></tr>
<tr><td>☐ spoon</td></tr>
<tr><td>☐ iron filings</td></tr>
<tr><td>☐ magnet</td></tr>
<tr><td>☐ sieve</td></tr>
<tr><td>☐ baking pan</td></tr>
</table>

Carry Out an Investigation

BE CAREFUL Wear safety goggles. Use caution to avoid spills.

1 Pour 1 cup of sand and half cup of salt into the bowl. Add a spoonful of iron filings and mix.

2 **Record Data** Use the magnet to remove the iron filings. Record your observations in the table on the next page.

3 Pour the water into the remaining mixture. Record your observations.

4 Pour the mixture through the sieve. Record your observations.

5 Pour the water into the shallow baking pan and allow it to evaporate. Record your observations in the table.

Tool	How well did it separate the mixture?	Which material did it separate best?
Magnet		
Sieve		
Evaporation		

Communicate Information

1. Did your results support your hypothesis? Explain why or why not.

_____

_____

_____

_____

2. Would any of the tools you used be effective at removing a liquid if you added liquid to the mixture? Explain.

_____

_____

_____

_____

Performance Task
Making Mixtures

You will make a variety of mixtures to show that mass is conserved when different types of matter are mixed. Then, you will try to identify the type of mixture that is made.

Make a Prediction Is mass conserved when two or more materials are mixed?

_____

_____

_____

_____

Carry Out an Investigation

BE CAREFUL Wear safety goggles. Use caution to avoid spills.

1 **Record Data** Measure 5 g of baking soda on a wax paper square and 20 milliliters (mL) of water in a graduated cylinder. Using a stirrer, mix them in one of the cups and measure the mass. Record the mass in the data table.

2 What type of mixture did you make? Record it in the table.

3 Repeat steps 1 and 2 with 5 grams (g) of baking soda and 5 g of baking powder.

4 Repeat steps 1 and 2 with 5 g of cornstarch and 20 mL of water.

5 Repeat steps 1 and 2 with 5 g of salt and 20 mL of vinegar.

6 Repeat steps 1 and 2 with 5 g of cornstarch and 5 g of salt.

7 Repeat steps 1 and 2 with 20 mL of water and 20 mL of vinegar.

Materials

- [ ] safety goggles
- [ ] 8 small, clear cups
- [ ] wax paper
- [ ] spoon
- [ ] 10 g baking soda
- [ ] pan balance
- [ ] 60 mL water
- [ ] graduated cylinder
- [ ] stirrers
- [ ] 10 g baking powder
- [ ] 40 mL vinegar
- [ ] 10 g cornstarch
- [ ] 10 g salt

Mixture	Mass or Volume	Type of Mixture
5 g Baking Soda + 20 mL Water		
5 g Baking Soda + 5 g Baking Powder		
5 g Cornstarch + 20 mL Water		
5 g Salt + 20 mL Vinegar		
5 g Cornstarch + 5 g Salt		
20 mL Water + 20 mL Vinegar		

Think like a pharmacist and investigate whether mixtures and solutions show conservation of mass.

Communicate Information

1. **Construct an Explanation** Was the mass of each mixture conserved? Why or why not?

_____

_____

_____

_____

_____

2. Did it matter what type of mixture was created for mass to be conserved? Explain.

_____

_____

_____

3. How could you show conservation of mass if you were mixing gases?

_____

_____

_____

_____

4. How could you graph the results of the investigation? Make your graph on a separate sheet of graph paper.

_____

_____

_____

_____

_____

Glue your graph here.

? Essential Question
What happens when different types of matter are mixed?

Think about the photo of a solid being mixed in water from the beginning of the lesson. Explain how the mass of individual materials in a mixture is conserved.

_____

_____

_____

_____

⚙ Science and Engineering Practices

Review the "I can . . ." statement you wrote earlier in the lesson. Explain what you have accomplished in this lesson by completing the "I did . . ." statement.

I did _____

_____

_____

_____

Now that you're done with the lesson, share what you did!

Chemical Changes

 **PAGE KEELEY SCIENCE PROBES** ## Chemical Change

frying an egg	tearing up paper	burning wood
a nail rusting	grilling a hamburger	dissolving salt in water
exploding fireworks	melting butter	baking a cake
rotting banana	mixing baking soda and vinegar	evaporating water
lighting a match	crushing a sugar cube	milk going sour

Changes in matter can be chemical or physical. Put an X in any of the boxes that are examples of chemical changes.

Explain your thinking. How did you decide if something is a chemical change?

_____

_____

_____

Science in Our World

▶ Watch the video of the candle burning. What is happening to the candle as it burns? What questions do you have?

_____

_____

_____

_____

Read about a chemical engineer and answer the questions on the next page.

A chemical engineer needs to know how matter interacts with other matter so he or she can use it to solve problems.

STEM Career Connection
Chemical Engineer

Do you ever think clothing could be made from old water bottles? Someday, it might! I am a chemical engineer, and I am investigating ways to recycle plastic from used water bottles into new fabrics that can be used to make things like clothing, furniture, and tents.

A chemical engineer solves problems that affect our everyday lives by applying the principles of chemistry. Many chemical engineers, like me, are interested in finding processes that convert raw materials or chemicals into more useful or valuable forms. Chemical engineers need to know how different types of matter interact so they can combine materials safely.

MALIK
Photonics Engineer

1. How might a chemical engineer think of ways to reuse plastic?

_____

_____

_____

2. What other materials do you think a chemical engineer would explore to find new ways of using them?

_____

_____

_____

? Essential Question

How does matter change when it interacts with other matter?

_____

_____

_____

⚙ Science and Engineering Practices

I will use mathematics and computational thinking.

Like a chemical engineer, you will use math to show conservation of mass in an investigation where the matter seems to change!

Inquiry Activity
Conservation of Mass

Is mass conserved even when matter reacts with other matter, such as when matter is mixed together and the matter appears to change?

Make a Prediction Will mass be conserved even when the matter seems to change into something new? Explain.

_____

_____

_____

_____

Carry Out an Investigation

BE CAREFUL Wear safety goggles when using the materials.

1. Measure 50 milliliters (mL) of water using the graduated cylinder and and pour it into the cup.

2. Measure 50 milliliters (mL) of water using the graduated cylinder and carefully pour it into the plastic bag. Keep the bag upright so the water does not spill. Seal the bag.

3. Measure 10 grams (g) of magnesium sulfate. Stir the magnesium sulfate into the cup of water.

4. **Record Data** Measure the mass of the magnesium sulfate solution. Record the mass in the table on the next page.

5. Add the sodium carbonate to the bag of water. Seal the bag and shake gently to stir.

6. **Record Data** Measure the mass of the sodium carbonate solution. Record the mass in the table.

7. Add the mass of each solution together. Record the total.

8. Open the bag, holding it upright. Carefully place the cup into the bag without spilling either solution. Reseal the bag.

9. Make sure the bag is sealed and tip the solution in the cup into the solution in the bag.

Materials
- [ ] safety goggles
- [ ] 100 mL water
- [ ] graduated cylinder
- [ ] small plastic cup
- [ ] resealable plastic bag
- [ ] 10 g magnesium sulfate
- [ ] spoon
- [ ] pan balance
- [ ] 10 g sodium carbonate

10 Measure and record the mass of the sealed bag, cup, and solutions after they are mixed.

	Mass in Grams
Mass of Magnesium Sulfate Solution	
Mass of Sodium Carbonate Solution	
Total Mass of the Solutions	
Mass of the Solutions After They Mix	

Communicate Information

1. What happened when the two solutions mixed together?

_____

_____

_____

2. Do you think this interaction could be reversed? Explain.

_____

_____

_____

3. Was mass conserved after the interaction? How do you know?

_____

_____

_____

4. Do you think the mass would have been conserved if the bag was open? Explain.

_____

_____

_____

Obtain and Communicate Information

abc Vocabulary

Use these words when explaining chemical changes.		
chemical properties	chemical change	chemical reaction
reactant	product	precipitate

Chemical Changes

Read pages 272–275 in the *Science Handbook.* Answer the
following questions after you have finished reading.

1. What is a chemical change also known as?

_____

2. Give an example from the text of a chemical change that you
cannot see.

_____

_____

3. What are reactants and products in a chemical reaction?

_____

_____

4. What other signs can a chemical change produce besides heat,
light, or gas?

_____

_____

_____

_____

FOLDABLES®

Cut out the Notebook Foldables tabs given to you by your teacher. Glue the anchor tabs as shown below. Use what you have learned to describe chemical changes in matter.

Glue anchor tab here

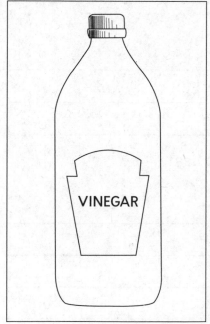

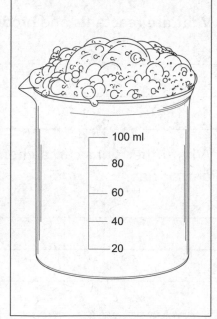

Types of Chemical Change

🖱 Explore the Digital Interactive *Types of Chemical Change* on the different kinds of common chemical reactions. Answer the questions when you have finished.

5. One type of chemical change is rust. What causes rust?

_____

_____

_____

6. What types of chemical changes occur in fireworks displays?

_____

_____

_____

_____

_____

⚙ # Science and Engineering Practices

Think about how you have used mathematics and computational thinking when exploring chemical changes. Tell how you can use math to describe conservation of mass in a chemical change by completing the "I can . . ." statement below.

Use examples from the lesson to explain what you can do!

I can _____

_____

_____

_____

Research, Investigate, and Communicate

 ## Inquiry Activity
Rate of Reaction

You will investigate whether the size of a solid that dissolves in water affects how quickly it will react with water.

Write a Hypothesis Will a whole or a crushed antacid tablet react faster with water? Write a hypothesis using an "If. . .then. . ." statement.

_____

_____

_____

Materials

- [ ] safety goggles
- [ ] 2 cups
- [ ] marker
- [ ] 200 mL of water
- [ ] 2 antacid tablets
- [ ] index card
- [ ] beaker

Carry Out an Investigation

BE CAREFUL Wear safety goggles. Use caution to avoid spills.

1. Label one of the cups Whole and the other Crushed.

2. Pour 100 mL of water into each of the cups.

3. Fold the index card in half. Place one antacid tablet inside the folded index card. Then, close the index card so that the antacid tablet is inside, and gently crush the antacid tablet with the bottom of the beaker. Be careful to not lose any of the pieces of the tablet.

4. **Record Data** At the same time, add the whole tablet to the Whole cup and the crushed tablet to the Crushed cup.

⚙ Crosscutting Concepts
Cause and Effect

Communicate Information

1. Which tablet had the stronger (faster) reaction?

_____

2. Was your prediction correct? Explain why or why not.

_____

_____

3. What other types of materials could you test the rate of their reaction with other matter?

_____

_____

_____

4. How do you think you could you slow down the rate of the reaction between the materials?

_____

_____

_____

Creating Chemistry

👁 Read *Creating Chemistry* about Antoine Lavoisier, the father of modern chemistry. Answer the questions when you have finished reading.

5. What big discoveries in chemistry is Lavoisier responsible for?

_____

_____

_____

_____

_____

_____

_____

6. What is considered Lavoisier's biggest accomplishment?

_____

_____

_____

_____

_____

_____

_____

_____

Performance Task
Changes in Matter

You will carry out an investigation in which you will observe a change from mixing substances together. You will then determine whether the mixture formed a new substance.

Make a Prediction Look at the materials list. What type of material do you think will be created in the investigation?

_____

_____

_____

Carry Out an Investigation

BE CAREFUL Wear safety goggles during the investigation.

1. Measure 1 teaspoon of borax and 125 mL of water in one of the beakers and stir until the borax dissolves.

2. **Record Data** Measure the mass of the solution and record the mass in the table.

3. Measure 60 mL of glue into the second beaker. Add the food coloring and stir until it is mixed.

4. **Record Data** Measure the mass of the glue and record the mass in the table.

5. Pour the borax solution into the glue. Stir well and observe.

6. **Record Data** Measure the combined mass of the beaker with the mixture and the empty beaker and record the total mass in the table.

Materials

- [ ] safety googles
- [ ] 2 250-mL beakers
- [ ] 125 mL warm water
- [ ] 1 teaspoon borax
- [ ] pan balance
- [ ] spoon
- [ ] 60 mL liquid school glue
- [ ] 5 drops of food coloring

Think like a chemical engineer and determine whether a mixture forms a new substance.

Mass of Beaker with Borax Solution	Mass of Beaker with Glue and Food Coloring	Mass of Both Beakers After Reaction

Online Content at connectED.mcgraw-hill.com

Communicate Information

1. What did you observe when materials from the two beakers were mixed?

_____

_____

_____

_____

2. How does the mass of both beakers after the reaction compare to the total mass of the individual beakers before the reaction?

_____

_____

_____

_____

3. **Construct an Explanation** Did the reaction show conservation of mass? Explain.

_____

_____

_____

_____

4. **Construct an Explanation** Was the result of the investigation a chemical change? Why or why not?

_____

_____

_____

_____

? **Essential Question**

How does matter change when it interacts with other matter?

▶ Think about the burning candle you saw at the beginning of the lesson. Explain what happens to matter in a chemical change.

_____

_____

_____

_____

⚙ **Science and Engineering Practices**

Now that you're done with the lesson, share what you did!

Review the "I can . . ." statement you wrote earlier in the lesson. Explain what you have accomplished in this lesson by completing the "I did . . ." statement.

I did _____

_____

_____

_____

Physical and Chemical Changes

⚙️ Performance Project
Plan Your Own Procedure

Ask a Question Use what you have learned about physical and chemical changes to plan your own investigation. You will show conservation of mass when sugar dissolves and when it melts. What question do you think your procedure will be able to answer?

_____

_____

_____

_____

Materials

☐ _____
☐ _____
☐ _____
☐ _____
☐ _____
☐ _____
☐ _____
☐ _____

Carry Out an Investigation

Write your procedure below. List the materials that someone will need to complete the investigation.

1 _____

_____

2 _____

_____

3 _____

_____

4 _____

_____

5 _____

How can an investigation show conservation of mass through change in matter?

Carry Out an Investigation

Follow the steps to your procedure to complete the investigation you planned, or trade procedures with a classmate and complete their procedure. Did the investigation answer the question you thought it would? Explain.

_____

_____

_____

_____

💬 How were the changes to matter different when it was dissolved versus when it was melted?

Make an Argument Does your procedure need to change to be successfully completed? How can you improve the procedure?

_____

_____

_____

_____

_____

_____

 ## Explore More in Our World

Did you learn the answers to all of your questions from the beginning of the module? If not, how could you design a different experiment or conduct research to help answer them?

Plant and Animal Needs

 ## Science in Our World

Plants and animals need certain things to survive. Look at the photo of the Venus flytrap. What do you wonder about these unique plants?

_____

_____

_____

_____

abc Key Vocabulary

> **Look and listen for these words as you learn about plant and animal needs.**
>
> aerobic respiration   anaerobic respiration   cell
>
> cellular respiration   chlorophyll   chloroplast
>
> energy   mitochondria   photosynthesis
>
> plant tropism   transpiration

How can I find out what materials are the most important for plants to grow?

OWEN
Entomologist

STEM Career Connection

Microbiologist

August 8th

The greenish tint to the water and the foam that stirred up when the waves hit the beach told us that there was something out of balance in this lake ecosystem. Algae provide the food base for most aquatic ecosystems and provide oxygen for the fish. But when there is too much algae, it can poison or choke other life-forms. Usually, algal blooms like this are caused by too many nutrients from fertilizers. Runoff from rainstorms takes the nutrients from the land to the lake. This causes the algae population to grow very quickly. As the excess algae die, the remains decay and clog the ecosystem. We'll take some samples of the water and look at it under the microscope to confirm our thoughts.

What did the microbiologist say caused the increase in algae in the lake?

_____

_____

_____

_____

Science and Engineering Practices

I will develop and use models.

I will engage in argument from evidence.

Name _____ Date _____

Plants and Photosynthesis

**PAGE KEELEY
SCIENCE
PROBES**

Plants and Food

Three friends were talking about plants. They each had different ideas about food and plants. This is what they said:

Amy: *Plants make their own food and they use the food they make.*

Mitch: *Plants make their own food but they don't use the food they make. They make the food for the animals that eat them.*

Beth: *Plants don't make their own food. They get the food they need from the soil.*

Who do you agree with most? _____

Explain why you agree.

_____

_____

_____

_____

Science in Our World

Look at the photo of the mangrove trees. How is this plant different from other plants you have seen? What do you wonder about these unique plants?

_____

_____

_____

_____

_____

Read about a plant scientist and answer the questions on the next page.

A plant scientist studies the processes that plants use to grow.

STEM Career Connection
Plant Scientist

As a plant scientist, I study ways to improve the quality of food plants by managing the crops of local farms. We use what we know to suggest the best crops to plant for the various climate conditions. Our work helps maintain the supply of produce across the country.

We try to breed plants so they are able to avoid disease and pests. I often work with other scientists to improve the breeding of bees and other insects that pollinate crops. If this is improved, then the crops have a better chance at reproducing, which will result in a bigger harvest of fruits and vegetables!

SAFFRON
Chef

1. How do plant scientists hope to improve plant processes?

_____

_____

_____

_____

2. How could you find out which resources are the most important for a plant to grow?

_____

_____

_____

? Essential Question

What do plants need to survive?

_____

_____

_____

 Science and Engineering Practices

I will engage in argument from evidence.

> Like a plant scientist, you will find out what is most important for plants to grow!

Inquiry Activity
Virtual Plant Simulation

What do plants need most to grow?

Investigate how the amount of water, sunlight, and carbon dioxide affect the growth of a virtual plant as you conduct the simulation.

Make a Prediction Which resource do you think will affect plant growth the most?

_____

_____

Carry Out an Investigation

1. How long does the virtual plant survive when you give it a high amount of light and carbon dioxide but no water?

_____

2. On one plant, set the light on high, the water on 20 mL, and the carbon dioxide on low. On the other plant, set the light on low, the water on 20 mL, and the carbon dioxide on low. What happens to the virtual plants?

_____

_____

3. Set all of the materials on one plant to the highest setting and on the second plant to the lowest setting. What happens?

_____

Communicate Information

1. How could you use the simulation to investigate which resource affected plant growth the most?

_____

_____

_____

Obtain and Communicate Information

🔤 Vocabulary

> **Use these words when explaining the needs of plants.**
>
> energy   cell   chloroplast
>
> chlorophyll   phloem   xylem
>
> photosynthesis   transpiration

Plant Needs

📖 Read pages 62–66 in the *Science Handbook.* Answer the following
   questions after you have finished reading.

1. Where must plants live in order to survive?

   _____

   _____

2. How do the xylem and phloem move water and food within
   the plant's transport system?

   _____

   _____

   _____

   _____

3. What is the process of transpiration?

   _____

   _____

   _____

Plant Cells

📖 Read pages 54–55 in the *Science Handbook.* Answer the following questions after you have finished reading.

4. What is able to pass through the cell wall of a plant cell?

_____

_____

_____

5. What role does the chloroplast play in a plant cell?

_____

_____

_____

_____

Photosynthesis

📖 Read pages 48–49 in the *Science Handbook.* Answer the following questions after you have finished reading.

⚙️ Crosscutting Concept
Energy and Matter

6. Why do plants need to perform photosynthesis?

_____

_____

_____

_____

_____

FOLDABLES®

Cut out the Notebook Foldables tabs given to you by your teacher. Glue the anchor tabs as shown below. Use what you have learned to describe the process of photosynthesis.

Glue anchor tab here

7. Explain the process of photosynthesis using words and a diagram. Use as much detail as possible and label your diagram.

_____

_____

_____

_____

_____

_____

Science and Engineering Practices

Think about how you have used evidence to argue what plants need to survive. Tell how you can use evidence to support an argument by completing the "I can . . ." statement below.

Use examples from the lesson to explain what you can do!

I can _____

_____

_____

Research, Investigate, and Communicate

 ## Inquiry Activity
Plant Investigation

You are going to investigate which material affects the growth of the plant the most.

Write a Hypothesis Which material will affect plant growth the most? Write your hypothesis as an "If...then..." statement.

_____

_____

_____

Materials
☐ plant pots
☐ soil, gravel, sand
☐ water
☐ beaker
☐ medium clear trash bags
☐ seeds
☐ pan balance
☐ graduated cylinder

Carry Out an Investigation

Plan your investigation below. Remember to investigate only one resource in your investigation.

_____

_____

_____

_____

_____

Record Data Record the growth of your plant in the table.

	Plant A	Plant B	Plant C (if testing soil)
Starting Height in cm			
Growth in cm After 3 Days			
Growth in cm After 6 Days			
Growth in cm After 9 Days			

Performance Task
Solution for Survival

You have learned which things plants need to carry out photosynthesis. Share what you have learned by designing a poster or brochure that provides a solution of how to care for a plant in a household setting to ensure that it gets everything it needs to survive.

Design a Solution Choose a plant to research and find out which things the plant needs to survive. Use the space below to plan your poster or brochure. Use a separate sheet of paper or other materials to create what you planned. After you create your poster or brochure, answer the questions on the next page.

Think like a plant scientist and share the best ways to take care of plants using evidence from the lesson.

Communicate Information

1. What type of plant and what kind of setting did you design your solution for?

_____

_____

_____

_____

2. Which resources does the plant need to survive?

_____

_____

_____

_____

_____

_____

3. How can you tell whether the plant is performing photosynthesis?

_____

_____

_____

_____

_____

_____

? Essential Question
What do plants need to survive?

Think about the mangrove trees you saw at the beginning of the lesson. Explain what plants need to survive.

_____

_____

_____

_____

_____

_____

⚙ Science and Engineering Practices

Review the "I can . . ." statement you wrote earlier in the lesson. Explain what you have accomplished in this lesson by completing the "I did . . ." statement.

Now that you're done with the lesson, share what you did!

I did _____

_____

_____

_____

Animals and Cellular Respiration

**PAGE KEELEY
SCIENCE
PROBES**

Why Do Animals Need Food?

Four friends were talking about animals and the food they need to survive. They each had different ideas. This is what they said:

Molly: *I think animals need food to get energy they need to live.*

Patty: *I think animals need food for growth and body repair.*

Devon: *I think animals need food for growth, body repair, and energy.*

Fidel: *I think animals need food so they won't be hungry.*

Who do you agree with most? _____

Explain why you agree.

_____

_____

_____

_____

Science in Our World

Look at the photo of the chipmunk. Why do animals need food?
What do you wonder about this animal's behavior?

_____

_____

_____

_____

_____

Read about a wildlife biologist and answer
the questions on the next page.

> A wildlife biologist might use models to track changes in the population of a plant or animal species.

STEM Career Connection
Wildlife Biologist

   I am a wildlife biologist for the United States
Fish and Wildlife Service, or FWS. I help research
and monitor habitats in Yellowstone National Park.
Lately, I have been studying some of the
predator-prey relationships within the park. In
1995, we introduced 41 wolves into the park after
they had been hunted to near extinction. Twenty
years later, we have more than 450 wolves. My
job is to find out what they eat to survive. It turns
out that during the summer they eat mostly deer,
rabbits, and other small mammals, but in the winter,
packs of wolves work together to take down the
much larger elk.

SAFFRON
Chef

1. What did the wildlife biologist find out about what wolves eat?

_____

_____

_____

_____

2. What is the wildlife biologist studying about the wolves?

_____

_____

_____

_____

? Essential Question

How do animals get energy from food?

_____

_____

_____

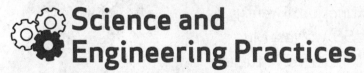

Science and Engineering Practices

I will develop and use models.

Like a wildlife biologist, you will make a model to show how animals get energy from food!

Inquiry Activity
Food and Respiration

How are different types of food broken down to be used as energy?

Make a Prediction Which of these liquids has the most sugar – juice, milk, water, regular soda, or diet soda?

_____

_____

_____

_____

Carry Out an Investigation

BE CAREFUL Wear safety goggles. Use caution to avoid spills.

1. Your teacher will provide 5 cups with a yeast solution. Use masking tape to label each cup with a testing liquid: juice, milk, water, regular soda, and diet soda.

2. Measure 120 mL of each of the testing liquids into the graduated cylinder. Carefully pour the testing liquids into the matching labeled cup. Rinse after each measurement.

3. **Record Data** Wait a few minutes and record your observations for each liquid.

4. After the yeast stops bubbling, rank the liquids based on the amount they bubbled, with 1 being the least amount of bubbles and 5 being the most amount of bubbles.

Materials

- [ ] safety goggles
- [ ] 5 clear plastic cups
- [ ] packages of active dry yeast
- [ ] 1 liter warm water
- [ ] masking tape
- [ ] marker
- [ ] graduated cylinder
- [ ] 120 mL of water
- [ ] 120 mL of apple juice
- [ ] 120 mL of milk
- [ ] 120 mL of regular soda
- [ ] 120 mL of diet soda

Type of Liquid	Observations	Amount of Bubbles

Communicate Information

1. How can you tell that the yeast is breaking down the liquids?

_____

_____

_____

2. Which liquid was broken down the most by the yeast? Why?

_____

_____

_____

_____

3. Which liquid was broken down the least by the yeast?

_____

_____

_____

4. How do you think this investigation relates to animals and the food they eat?

_____

_____

_____

_____

_____

_____

Obtain and Communicate Information

abc Vocabulary

> Use these words when explaining how animals get and
> use energy from food.
>
> cellular respiration   mitochondria
>
> aerobic respiration   anaerobic respiration

Animals Using Energy

▶ Watch *Animals Using Energy.* Answer the following
   questions when you have finished watching.

1. Do all animals need to eat the same amount of food to get energy?
   Why or why not?

_____

_____

_____

_____

_____

2. What types of foods do you eat for energy?

_____

_____

_____

Animals Use Cellular Respiration

📲 Explore the Digital Interactive *Animals Use Cellular Respiration* about how animals use food for energy. Answer the following questions after you have finished.

3. Where does the energy in photosynthesis and cellular respiration come from?

_____

_____

4. What part of an animal cell breaks down food for energy?

_____

_____

5. What types of things do animals use energy to do?

_____

_____

Animal Needs

📖 Read pages 72–73 in the *Science Handbook*. Answer the following questions after you have finished reading.

6. What are the main needs of animals to be able to survive?

_____

_____

_____

7. How do animals get energy to meet these needs?

_____

_____

_____

_____

Cellular Respiration and Energy from Food

📖 Read pages 50–51 in the *Science Handbook.* Answer the following questions after you have finished reading.

⚙️ Crosscutting Concepts
Energy and Matter

8. Besides energy, what else is produced during cellular respiration?

_____

_____

_____

9. What are mitochondria?

_____

_____

_____

10. What are nutrients? Give three examples of different types of nutrients.

_____

_____

_____

_____

_____

Animal Cells

📖 Read pages 52–53 in the *Science Handbook*. Answer
the following questions after you have finished reading.

11. Why do some types of animal cells have more mitochondria?

_____

_____

_____

_____

12. What is one main difference between plant cells and
animal cells?

_____

_____

_____

Science and Engineering Practices

Think about how you could use a model to show what
you have learned about how animals get energy from
food. Tell how you can develop and use models
by completing the "I can . . ." statement below.

Use examples from
the lesson to explain
what you can do!

I can _____

_____

_____

_____

_____

_____

Research, Investigate, and Communicate

Types of Respiration

Read *Types of Respiration* on the two types of cellular respiration that can occur. Answer the following questions after you have finished reading.

1. What is the difference between aerobic and anaerobic respiration?

_____

_____

_____

_____

2. What types of living things use anaerobic respiration?

_____

_____

_____

_____

3. Describe the chemical formula of cellular respiration.

_____

_____

_____

_____

_____

FOLDABLES®

Cut out the Notebook Foldables tabs given to you by your teacher. Glue the anchor tabs as shown below. Use what you have learned to give an example of each term and describe how it relates to energy. Which is more efficient? Write your answers under the tabs.

Glue anchor tab here

⚙ Performance Task
Comparing Cells

You will make a model of a plant cell and an animal cell by defining the structures and comparing and contrasting their parts. You will use evidence from the lesson to support the argument that the differences in cells lead to plants and animals getting energy through different processes.

Make a Model Use the boxes below to make your model of a plant cell and an animal cell. Make sure you label your diagrams and explain the role of each cell part.

Communicate Information

1. How does the structure of a plant cell support photosynthesis?

_____

_____

_____

_____

_____

2. How does the structure of an animal cell support cellular respiration?

_____

_____

_____

_____

_____

3. What is similar between plant and animal cells?

_____

_____

_____

4. What is different between plant and animal cells?

_____

_____

_____

_____

_____

_____

? Essential Question
How do animals get energy from food?

Think about the chipmunk hoarding food that you saw at the beginning of the lesson. Explain how animals get energy from food and how they use that energy.

_____

_____

_____

_____

⚙ Science and Engineering Practices

Review the "I can..." statement you wrote earlier in the lesson. Explain what you have accomplished in this lesson by completing the "I did.." statement.

Now that you're done with the lesson, share what you did!

I did _____

_____

_____

Name _____ Date _____

Plants and Cellular Respiration

Plant and Animal Processes

The chart below lists 3 processes carried out by living organisms. Put an X in the boxes to show which processes are carried out by plants, animals, or both plants and animals.

Process	Plants Only	Animals Only	Plants and Animals
Growth			
Photosynthesis			
Cellular Respiration			

Explain your thinking. How did you decide which processes are carried out by animals only, plants only, or both plants and animals?

_____

_____

_____

_____

Science in Our World

Look at the photos of the ivy and sunflowers. Do they both need the same things to survive? What do you wonder about the way these plants are growing?

_____

_____

_____

_____

Read about an agricultural engineer and answer the questions on the next page.

An agricultural engineer might make a model of new farm equipment to make the task more efficient.

STEM Career Connection
Agricultural Engineer

As an agricultural engineer, I work with the farming community to improve production, product quality, and the environment for farmworkers and animals. Last year, I designed a water reservoir for a lettuce farmer who needed to conserve water.

When I survey a farm's production facilities, I ask myself, "How can this job be done more efficiently?" Agricultural engineers develop methods and design equipment for land preparation, planting and harvesting. They also improve upon existing farming technologies.

SAFFRON
Chef

1. Which elements of farming does an agricultural engineer work to improve?

_____

_____

_____

_____

2. What do you think an agricultural engineer needs to consider about the needs of plants?

_____

_____

_____

_____

_____

? Essential Question
How do plants use energy?

_____

_____

_____

⚙ Science and Engineering Practices

I will develop and use models.

Like an agricultural engineer, you will consider the needs of plants to develop a model!

Inquiry Activity
Virtual Plant Simulation Revisit

Does the amount of each resource a plant needs affect its growth?

Investigate how much sugar the plant is able to produce relative to the amount of resources it gets as you conduct the simulation.

Make a Prediction Which resource on its highest setting will result in the highest sugar production in the plant?

_____

_____

Carry Out an Investigation

1. On one plant, set the light on low, the water on 20 mL, and the CO_2 on low. On the other plant, set the light on high, the water on 20 mL, and the CO_2 on low. Which plant made more sugar?

_____

_____

2. On one plant, set the light on low, the water on 20 mL, and the CO_2 on low. On the other plant, set the light on low, the water on 40 mL, and the CO_2 on low. Which plant produced more sugar?

_____

_____

3. On one plant, set the light on low, the water on 20 mL, and the CO_2 on low. On the other plant, set the light on low, the water on 20 mL, and the CO_2 on high. Which plant produced more sugar?

_____

_____

Communicate Information

1. Which resource had the greatest effect on sugar production? Was your prediction correct?

_____

_____

Obtain and Communicate Information

abc Vocabulary

> **Use this word when explaining how plants use energy from food.**
>
> plant tropism

Plants Use Cellular Respiration

👁 Read *Plants Use Cellular Respiration* on how plants use the food they make during photosynthesis. Answer the following questions when you have finished reading.

1. What processes does a plant use to produce its own food for energy and to store that energy?

_____

_____

_____

_____

2. Where does cellular respiration occur in plant cells?

_____

_____

3. What do plants use the energy from cellular respiration to do?

_____

_____

_____

_____

Inquiry Activity
Area of a Leaf

The amount of sugar and oxygen that a leaf produces is related to the leaf's surface area. You will find the surface area of different-sized leaves and make inferences about which one produces more sugar.

Materials

☐ 2 leaves from different plants

☐ graph paper

☐ pencil

Carry Out an Investigation

Record Data Find two different-sized leaves from different types of trees. Follow the steps below and record the data in the table to find the area of each leaf.

1. Trace the leaves on graph paper.

2. Count the number of whole square units.

3. Count the number of partial square units and divide by 2.

4. Add the two numbers for the total surface area of the leaf.

	Whole Squares +	Partial Squares ÷ 2	= Area
Leaf 1			
Leaf 2			

Communicate Information

4. Which leaf produces more sugar and oxygen? How do you know?

_____

_____

_____

5. If larger leaves produce more food for the tree, why do you think some trees benefit from smaller leaves or tiny needles?

_____

_____

FOLDABLES®

Cut out the Notebook Foldables tabs given to you by your teacher.
Glue the anchor tabs as shown below. Use what you have learned to
explain how a plant gets energy for new growth.

Glue anchor tab here

Plant Investigation

Revisit your *Plant Investigation* data that you have been collecting since Lesson 1. Answer the following questions to analyze the results of your investigation.

6. How did withholding one resource affect the growth of your plant?

_____

_____

_____

_____

7. Compare your results with the results of your classmates. Which material affects plant growth the most?

_____

_____

_____

_____

_____

Science and Engineering Practices

Think about how you have developed and used models to understand how plants carry out photosynthesis and cellular respiration. Tell how you can use a model to explain how plants get and use energy by completing the "I can . . ." statement below.

Use examples from the lesson to explain what you can do!

I can _____

_____

_____

_____

Research, Investigate, and Communicate

Plant Tropisms

Although plants cannot move like animals, some plants react to their environment. This response is known as a *tropism*. The four main types are *phototropism*, *gravitropism*, *hydrotropism*, and *thigmotropism*.

📖 Use your *Science Handbook* on page 46 and other research materials to find information about one of the main types of plant tropisms. Record your research below.

_____

_____

_____

_____

_____

_____

_____

Use your research to draw a diagram of the tropism you researched. Make sure you label the components of the diagram.

Carry Out an Investigation

You are going to plan an investigation about the plant tropism using the information from your research.

Ask a Question What question do you hope to answer about the plant tropism from the results of the investigation you are going to plan?

_____

_____

_____

1 List the materials you will need to carry out the investigation.

_____

_____

_____

_____

_____

_____

2 How much time will you need in order to start seeing results? Explain.

_____

_____

_____

_____

_____

Carry Out an Investigation

Write the procedure that you will follow to carry out the investigation of the plant tropism you researched. Make sure your steps are clearly written so someone else could follow the procedure and replicate your results.

_____

_____

_____

_____

_____

_____

_____

_____

Crosscutting Concepts
Energy and Matter

1. How do tropisms help plants?

_____

_____

_____

_____

_____

_____

_____

_____

⚙ Performance Task
How Plants Use Energy

You will create a model of a plant that shows how cellular respiration and photosynthesis help plants survive. Be sure your model shows the source of energy, resources the plant needs, and how these resources enter and leave the different parts of the plant.

Make a Model Decide which materials you want to use to make your model. Use the space below to sketch your model, and then build the model using your sketch. Answer the questions on the next page when you are done. Share your model with your classmates.

Think like an agricultural engineer and show what you know about how plants get and use energy!

Communicate Information

1. What is the source of all energy for the plant?

_____

_____

2. What other types of matter did you show entering and leaving parts of the plant?

_____

_____

_____

_____

_____

_____

_____

3. How did you show the differences between photosynthesis and cellular respiration in your model? How are they related?

_____

_____

_____

_____

_____

_____

_____

_____

? Essential Question
How do plants use energy?

Think about the ivy and sunflowers from the beginning of the lesson. Explain how these plants are getting and using energy to survive.

_____

_____

_____

_____

_____

⚙ Science and Engineering Practices

Review the "I can…" statement that you wrote earlier in the lesson. Explain what you have accomplished in this lesson by completing the "I did…" statement.

Now that you're done with the lesson, share what you did!

I did _____

_____

_____

_____

_____

_____

_____

Plant and Animal Needs

⚙ Performance Project
With or Without Soil?

Research hydroponic plants and find out whether or not plants can get what they need to grow without soil. Write your notes here.

_____

_____

_____

_____

_____

Plan an investigation to study hydroponic plants yourself. Make sure you list the materials and procedure you will need.

_____

_____

_____

_____

_____

_____

How can I find out what materials are the most important for plants to grow?

Do plants need soil to grow? Explain.

_____

_____

_____

_____

_____

_____

_____

_____

 What resources do plants need to grow? Can all plants grow without soil? Talk about what regions hydroponic plants can grow.

● Explore More in Our World

Did you learn the answers to all of your questions from the beginning of the module? If not, how could you design a different experiment or conduct research to help answer them?

Matter in Ecosystems

 ## Science in Our World

Look at the photo of compost. What is the compost made of?
What do you wonder about the compost and how it is used?

_____

_____

_____

_____

_____

_____

_____

_____

abc Key Vocabulary

Look and listen for these words as you learn about matter in ecosystems.

abiotic factor	biotic factor	food chain
food web	invasive species	limiting factor
nitrogen cycle	oxygen-carbon dioxide cycle	succession
water cycle		

What can a compost pile help ecologists understand about ecosystems?

HIRO
Ocean Engineer

STEM Career Connection
Ecologist

As an ecologist, I have the important job of studying and understanding nature. I have seen how living things are connected to each other when they live in the same or nearby ecosystems. Sometimes I work with environmental organizations to manage or create habitats when I observe a need for a struggling animal or plant population. My favorite thing to study is conservation.

Compost is a great example of how we can recycle waste. The parts of fruits and vegetables that we don't eat can be reused for so much more than just throwing them away! The materials in compost give plants the nutrients they need to grow. Wait a minute, I think I see worms, flies, and other insects using the compost, too!

How do ecologists help living things in ecosystems?

_____

_____

_____

_____

 **Science and Engineering Practices**

I will develop and use models.

Interactions of Living Things

Energy and Matter in Ecosystems

Three friends were talking about how matter and energy move through living things in an ecosystem. They each had different ideas. This is what they said:

Ted: I think matter gets recycled in ecosystems, but energy does not get recycled.

Jill: I think energy gets recycled in ecosystems, but matter does not get recycled.

Lee: I think both matter and energy get recycled in ecosystems.

Who do you agree with most? _____

Explain why you agree.

_____

_____

_____

_____

Science in Our World

Look at the photo of the predator and prey. How does the predator affect the population of the prey? What do you wonder about how these living things interact?

_____

_____

_____

_____

Read about an environmental scientist and answer the questions on the next page.

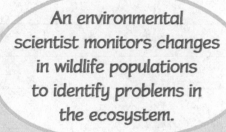

> An environmental scientist monitors changes in wildlife populations to identify problems in the ecosystem.

STEM Career Connection
Environmental Scientist

June 10th

After testing the water supply, we have found high levels of pesticides. Since there is a major corn farm in this region, we need to monitor the levels of pesticides over the next few days to see if they improve. The plants and animals in this pond ecosystem seem to be at healthy population levels.

July 2nd

The levels of pesticides in the water have increased. With minimal amounts of precipitation over the past couple of weeks, the water has not been able to stabilize. We are concerned about wildlife populations, because the number of herons has decreased, and we have found frog eggs that did not hatch.

POPPY
Park Ranger

1. What types of events does an environmental scientist study in ecosystems?

_____

_____

_____

_____

2. How could the environmental scientist help solve the pesticide problem?

_____

_____

_____

_____

? Essential Question

How does energy flow in an ecosystem?

_____

_____

⚙ Science and Engineering Practices

I will develop and use models.

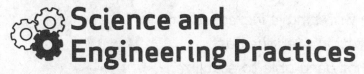

Like an environmental scientist, you will use models to show how living things interact in an ecosystem!

Inquiry Activity
Foxes and Rabbits

Materials

☐ masking tape

☐ 8 7.5-cm cardboard squares

☐ 100 2.5-cm construction paper squares

How do predator and prey relationships affect each other?

Make a Prediction What happens to the population of rabbits when the population of foxes increases?

_____

_____

_____

_____

_____

Carry Out an Investigation

1. Use the tape to mark off a 60-cm square. This square represents a forest. Distribute 10 of the small squares within the forest. These squares represent rabbits.

2. The larger squares represent foxes. The fox must touch at least one rabbit square to live. If it touches three or more rabbits, then it will reproduce. If the fox reproduces, then you will toss another fox in for the next trial.

3. **Record Data** Toss one fox into the forest. Remove any rabbits that the fox touches. Record the results in the data table on the next page.

4. At the start of the next trial, double the number of rabbits remaining from the first trial to represent new rabbit offspring. Place these new rabbits in the forest.

5. If the entire rabbit population was removed by the fox, add three new rabbits to the forest to represent new rabbits moving into the area. If all of your foxes starve, then add a fox to represent a new fox moving into the area.

6. In each additional trial, throw each fox square once. This includes any surviving foxes from previous trials and any offspring produced in previous trials. Record the results in the data table.

Trial	Number of Rabbits Left	Number of Foxes Left	Number of Rabbits Caught	Number of New Rabbits in Next Trial	Number of New Foxes in Next Trial

Communicate Information

1. What happened to the population of foxes as the population of rabbits increased?

_____

_____

2. Was your prediction correct? Why or why not?

_____

_____

3. What other populations in a forest ecosystem might be affected by these population changes?

_____

_____

_____

4. What would happen if the plant population in the forest decreased?

_____

_____

Obtain and Communicate Information

abc Vocabulary

> Use these words when explaining the interactions of living things in ecosystems.
>
> biotic factor   abiotic factor   niche
>
> food chain   food web   consumer
>
> producer   decomposer

Ecosystems

📖 Read pages 88–89 in the *Science Handbook*. Answer the following questions after you have finished reading.

1. What is the difference between biotic and abiotic factors?

_____

_____

_____

_____

2. What is a community in an ecosystem?

_____

_____

_____

3. Name three biotic factors in a forest ecosystem.

_____

_____

_____

FOLDABLES®

Cut out the Notebook Foldables tabs given to you by your
teacher. Glue the anchor tabs as shown below. Use what
you know to describe the different types of living things
in ecosystems.

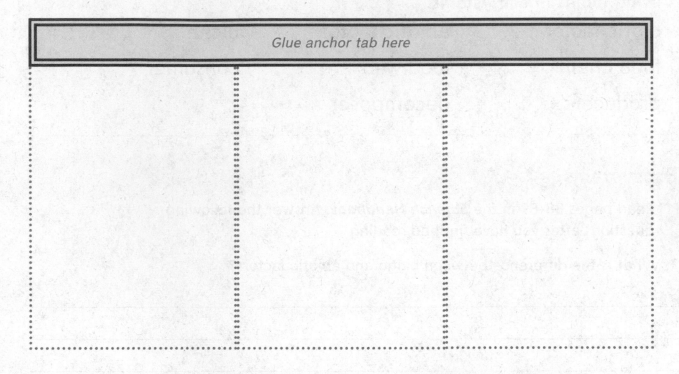

Glue anchor tab here

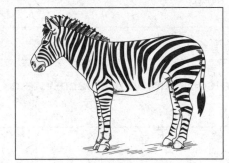

Food Chains and Food Webs

📖 Read pages 102–103 in the *Science Handbook.* Answer
   the following questions after you have finished reading.

⚙ Crosscutting Concepts
Systems and System Models

4. How do food chains show energy moving through an ecosystem?

_____

_____

_____

5. How would a food web be affected if a food source for several
   different predators died out?

_____

_____

_____

_____

Energy Flow in a Food Chain

🔎 Explore the Digital Interactive *Energy Flow in a Food Chain* about
   the relationship of some living things in their ecosystem. Answer the
   questions after you have finished.

6. What do each of the biotic factors in the activity need to do?

_____

_____

7. What would happen if the grasshoppers were removed from
   the food chain?

_____

_____

_____

Model of a Food Chain

Research Choose an ecosystem to research. Use the table below to record information about the plants and animals that live in the ecosystem you chose.

Type of Ecosystem	
_____	
Plants	**Animals**

8. Where do the plants get their energy?

_____

_____

9. What animals eat plants? What animals eat other animals?

_____

_____

Use the information you collected to plan your model of the food chain. Draw a sketch of the flow of matter and energy between the living things on a separate sheet of paper. Remember to show the source of energy in the ecosystem.

Use examples from the lesson to explain what you can do!

⚙ Science and Engineering Practices

Think about how you have learned about and created a food chain. Tell how you can develop and use models by completing the "I can . . ." statement below.

I can _____

_____

🔍 Research, Investigate, and Communicate

Energy in a Salt Marsh Ecosystem

You have learned how living things in ecosystems interact with each other and how these interactions can be modeled with food chains and food webs.

👁 Read *Energy in a Salt Marsh Ecosystem*. Use the space below to draw two different food chains from the reading.

Energy in a Salt Marsh Ecosystem

Look at your models of salt marsh food chains. Do they connect to make a food web? In the space below, show how the two food chains interact.

⚙ Performance Task
Build a Food Web

You will think like an environmental scientist to make a model of a food web using what you have learned about the interactions of living things in an ecosystem.

Make a Model Choose an ecosystem. Research information about the living things in the ecosystem and plan a model of a food web below to show how the living things interact. Use available materials to build your model using your plan.

Think like an environmental scientist and make a model of a food web!

Communicate Information

1. Explain how your food web model shows transfer of energy between living things in an ecosystem.

_____

_____

_____

_____

_____

2. How do the abiotic and biotic factors interact in the ecosystem?

_____

_____

_____

_____

3. Think about a situation where the plants in your model become diseased. What kind of solution could you propose to fix this problem?

_____

_____

_____

_____

_____

? Essential Question
How does energy flow in an ecosystem?

Think about the photo of a predator-prey relationship from the beginning of the lesson. Explain how prey is affected by an increase in predators, and how a predator is affected by a decrease in prey.

_____

_____

_____

_____

_____

_____

_____

_____

Science and Engineering Practices

Review the "I can..." statement you wrote earlier in the lesson. Explain what you have accomplished in this lesson by completing the "I did..." statement.

I did _____

_____

_____

_____

Now that you're done with the lesson, share what you did!

Balance in Ecosystems

PAGE KEELEY
SCIENCE
PROBES

Let It Go?

Liam showed his friend the fish in his aquarium. One of his fish has grown too big for the aquarium. Liam and his friend disagreed about what should be done with the big fish. This is what they said:

Liam: *I think I should release it into the wild. It will survive because our climate is the same as the one the fish came from.*

Betsy: *I don't think you should release it into the wild. You will have to figure out something else to do for the fish.*

Who do you agree with most? _____

Explain why you agree.

_____

_____

_____

Science in Our World

Look at the photo of a damaged tree. Will the tree be able to survive? What do you wonder about how this happened?

_____

_____

_____

_____

Read about a wildlife conservationist and answer the questions on the next page.

> A wildlife conservationist tries to find solutions to problems that damage the balance in ecosystems.

STEM Career Connection
Wildlife Conservationist

April 18th

After a survey of the local forest, we noticed a decrease in the number of butterflies and decided to investigate the cause. The butterfly population has been strong in recent years, and they are necessary to help pollinate the various flowers and plants in the area.

April 30th

We have discovered an overwhelming amount of butterfly bushes in the area. These plants are not native to the area, and they actually disrupt the life cycle of the butterfly. Although adult butterflies can use the nectar from these plants, their larvae cannot be sustained, and they end up dying. We are in the process of finding a solution to limit the spread of these butterfly bushes.

POPPY
Park Ranger

Name _____ Date _____

1. What problem did the wildlife conservationist identify?

_____

_____

_____

_____

2. What solution do you think the wildlife conservationist can suggest to help improve the butterfly population?

_____

_____

_____

_____

? Essential Question
How do changes affect ecosystems?

_____

_____

_____

_____

Science and Engineering Practices

I will develop and use models.

> Like a wildlife conservationist, you will use models to show how balance is maintained in an ecosystem!

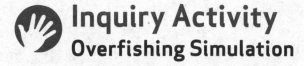

Inquiry Activity
Overfishing Simulation

How can overfishing affect fish populations over time?

▦ Investigate how the balance in a lake ecosystem can be affected by overfishing as you conduct the simulation.

Make a Prediction How will the number of fish caught relate to the number of boats used for fishing?

_____

_____

_____

_____

_____

_____

Carry Out an Investigation

1 For each of the first five fishing seasons, choose only one boat to send out to fish in the lake. How many total fish were caught?

_____

_____

2 For each of the next five fishing seasons, choose two boats to send out to fish in the lake. How many total fish were caught?

_____

_____

3 For each of the next five fishing seasons, choose three boats to send out to fish in the lake. How many total fish were caught?

_____

_____

Communicate Information

1. How did the fishing boats affect the population of fish in the lake ecosystem?

_____

_____

_____

_____

2. What number of boats resulted in the greatest number of fish caught? Do you think this could last?

_____

_____

_____

_____

3. What other factors might affect the fish population? Consider factors that did not appear in the model.

_____

_____

_____

_____

Obtain and Communicate Information

🔤 Vocabulary

> **Use these words when explaining balance in ecosystems.**
>
> limiting factor   symbiosis   mutualism
>
> commensalism   parasitism   invasive species

Resources in Ecosystems

📖 Read page 97 in the *Science Handbook*. Answer the following questions after you have finished reading.

1. What happens when there is a limited amount of resources in an ecosystem?

_____

_____

_____

_____

2. Give an example of an abiotic limiting factor. How does it affect the balance in ecosystems?

_____

_____

_____

_____

_____

_____

Name _____ Date _____

 **Inquiry Activity**
Limiting Plant Growth

Living and nonliving things affect the balance in an ecosystem. You will investigate how limiting factors affect plant growth.

Write a Hypothesis Does space affect plant growth? Write a hypothesis in the form of an "If...then..." statement.

_____

_____

_____

Carry Out an Investigation

1. Use the small squares from Lesson 1. Each small square represents the space that a plant needs to grow. Cut out a 20-cm square to represent the environment.

2. **Record Data** Place the 20-cm square in front of you. Pick up 8 small squares and toss them into the environment. Rethrow any squares that do not land in the environment. Remove squares that are touching. If a square lands in the environment without touching another square, it survives. Record the number of plants that have survived.

3. Repeat step 2 six more times, but increase the number of plants by two each time. Record your results.

Number of Plants Tossed in Trial	Number of Plants That Survive
8	
10	
12	
14	
16	
18	

4 **Analyze Data** Use the data you recorded in the activity to create a graph of the plant population at each trial. Label your graph.

Communicate Information

3. How is the space that plants have to grow a limiting factor?

_____

_____

_____

_____

_____

_____

_____

4. How could you solve for this limiting factor to increase the number of plants that can grow?

_____

_____

_____

_____

_____

_____

_____

Glue your graph here.

Habitats and Niches

▶ Watch *Habitats and Niches* on the role of living things in ecosystems. Answer the questions when you have finished watching.

5. What niche does a kelp forest fulfill in an ocean ecosystem?

_____

_____

_____

6. How do the herbivores in a savanna habitat avoid competition?

_____

_____

_____

_____

_____

Niches Reduce Competition

Explore the Digital Interactive *Niches Reduce Competition* on the effect of niches. Answer the question when you have finished.

7. How do the niches of the cedar waxwing and the cardinal help them avoid competition?

_____

_____

_____

_____

_____

_____

Symbiosis and Invasive Species

📖 Read page 119 in the *Science Handbook.*

👁 Then, read *Invasive Species* about organisms that disrupt the balance in ecosystems. Answer the following questions after you have finished reading.

⚙ Crosscutting Concepts
Systems and System Models

8. How are mutualism, commensalism, and parasitism the same? How are they different?

_____

_____

_____

_____

_____

_____

_____

_____

9. What did all of the examples of invasive species have in common?

_____

_____

_____

Name _____ Date _____

FOLDABLES®

Cut out the Notebook Foldables tabs given to you by your teacher. Glue the anchor tabs as shown below. Use what you have learned to describe factors that affect balance in ecosystems.

Glue anchor tab here

Glue anchor tab here

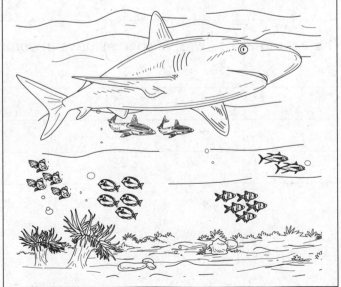

Changes in Ecosystems

📖 Read pages 104–107 in the *Science Handbook*. Answer the following questions when you have finished reading.

10. What is a reason that populations of organisms might move between ecosystems?

_____

_____

_____

_____

11. How can natural events change the balance in an ecosystem?

_____

_____

_____

_____

⚙️ Science and Engineering Practices

Think about how you have used models to show the balance between living and nonliving things in an ecosystem. Tell how you can develop and use models by completing the "I can..." statement below.

Use examples from the lesson to explain what you can do!

I can _____

_____

_____

Research, Investigate, and Communicate

Overfishing Simulation Revisit

Investigate how you can best maintain the balance of the fish population in the lake ecosystem, rather than trying to catch the most fish, as you revisit the simulation.

Carry Out an Investigation

1 **Record Data** Sketch the graph of the balanced lake ecosystem below. What number of fishing boats did you send out each season to keep the graph of the total fish population in the lake the most balanced?

_____

_____

_____

_____

_____

Performance Task
Solve for an Invasive Species

You will think like a wildlife conservationist to research an invasive species, define a problem it causes, and design a possible solution to the problem.

Research Find information about an invasive species that interests you. Record your notes below.

1 What is the name of the invasive species?

_____

2 Where is the invasive species originally from?

_____

3 Which ecosystem is it in now? How did it get there?

_____

_____

_____

4 How does this species affect the ecosystem it is in now?

_____

_____

_____

5 What other information is interesting about this species?

_____

_____

Define a Problem Use your research to define the problem that is caused by the invasive species.

_____

_____

_____

Design a Solution

Use what you have learned about balance in ecosystems to design a model of a solution to the problem caused by the invasive species that you researched.

Communicate Information

1. How does your solution solve the problem that is caused by the invasive species?

_____

_____

2. What can others learn from your solution in order to help stop the spread of invasive species?

_____

_____

? Essential Question
How do changes affect ecosystems?

Think about the photo of the damage to the tree that you saw at the beginning of the lesson. Explain how an invasive species is affecting the population of the trees.

_____

_____

_____

_____

_____

_____

⚙ Science and Engineering Practices

Review the "I can..." statement you wrote earlier in the lesson. Explain what you have accomplished in this lesson by completing the "I did..." statement.

Now that you're done with the lesson, share what you did!

I did _____

_____

_____

_____

_____

Name _____ Date _____

Cycles in Ecosystems

What Happens to the Matter?

Two friends disagreed about what happens to the matter in an ecosystem as organisms die and living organisms eat, breathe, and drink water. This is what they said:

Baye: *I think the matter gets recycled in the living and nonliving things in an ecosystem.*

Arlen: *I think the matter eventually gets used up and then is gone from the ecosystem.*

Who do you agree with the most? _____

Explain why you agree.

_____

_____

_____

_____

Science in Our World

Look at the photo of the greenhouse. How are plants able to grow in this structure? What do you wonder about the greenhouse?

_____

_____

_____

_____

Read about a botanist and answer the questions on the next page.

A botanist needs to consider the types of materials that cycle through plants.

STEM Career Connection
Botanist

What a beautiful day to be working on the farm! I'm a botanist, and I work for the United States Department of Agriculture, or USDA. A botanist studies plant life, from the tiniest microorganism to the tallest trees. I work with the types of plants grown on farms and used for food. I help food growers provide the healthiest food for you and your family to eat.

I work with farmers to develop the safest means of pest control and to limit the use of pesticides that may be harmful. I also help farmers grow their crops more efficiently and with less water. Another part of my job is to develop hybrid plants that provide more food per plant or need less care than other varieties.

POPPY
Park Ranger

1. What does the botanist's journal entry tell you about the job of a botanist?

_____

_____

_____

2. What questions do you have about the role of botanists?

_____

_____

_____

? Essential Question
How is matter cycled through ecosystems?

_____

_____

_____

_____

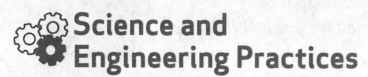 # Science and Engineering Practices

I will develop and use models.

Like a botanist, you will consider the cycles in ecosystems to make a model!

Inquiry Activity
Cycling Matter

How does water cycle through an ecosystem?

Make a Prediction How will the water change state to cycle through the system?

_____

_____

_____

Materials

☐ clear container

☐ warm water

☐ plastic wrap

☐ large rubber band

☐ marble

☐ ice cubes

Carry Out an Investigation

1. Fill the clear container ¼ full with warm water.

2. Stretch plastic wrap over the top of the container. Use a rubber band to secure the plastic wrap.

3. Place a marble and several ice cubes in the center of the top of the plastic wrap.

4. **Record Data** Check on the cups every 5 minutes and record your observations.

5 minutes:

_____

_____

10 minutes:

_____

_____

15 minutes:

_____

_____

20 minutes:

_____

_____

_____

Communicate Information

1. Where do the water droplets come from?

_____

_____

_____

_____

_____

2. What would happen if the container were not covered by
the plastic wrap?

_____

_____

_____

_____

_____

3. Would a change in the temperature of the water affect how the
water droplets form?

_____

_____

_____

_____

Obtain and Communicate Information

abc Vocabulary

Use these words when explaining how matter is cycled through an ecosystem.

water cycle   water vapor   oxygen-carbon
                                          dioxide cycle
nitrogen cycle

Terrariums

▶ Watch *Terrariums* on how matter is cycled in a closed system. Answer the following questions after you have finished watching.

1. What types of matter are being cycled in the terrarium?

_____

2. How is water cycled through different parts of the terrarium?

_____

_____

_____

3. Why is water important to this ecosystem model?

_____

4. How is Earth like a terrarium?

_____

_____

_____

_____

The Water Cycle

📖 Read pages 186–187 in the *Science Handbook*. Answer
the following questions after you have finished reading.

6. What is transpiration?

_____

_____

7. What different things can happen to rainwater on land?

_____

_____

_____

_____

8. What is a watershed?

_____

_____

Oxygen-Carbon Dioxide Cycle

👁 Read *Oxygen-Carbon Dioxide Cycle* on how two important gases
cycle through ecosystems. Answer the questions after you have
finished reading.

.9. Describe the oxygen-carbon dioxide cycle.

_____

_____

_____

_____

10. What are other sources of carbon dioxide?

_____

_____

Energy in Ecosystems

📖 Read pages 98-99 in the *Science Handbook*. Answer the following questions after you have finished reading.

11. What are two things that can fix atmospheric nitrogen into a form that plants can use?

_____

_____

_____

12. What different forms does nitrogen take in the nitrogen cycle?

_____

_____

_____

13. How do plants get nitrogen in a form they can use?

_____

_____

_____

_____

14. How do animals return nitrogen back to the soil?

_____

_____

_____

_____

_____

Name _____ Date _____

FOLDABLES®

Cut out the Notebook Foldables tabs given to you by your teacher. Glue the anchor tabs as shown below. Use what you have learned to describe the three cycles in a rain forest.

Glue anchor tab here

Nitrogen Cycle

Explore the Digital Interactive *Nitrogen Cycle* on how the forms of nitrogen cycle through ecosystems. Answer the questions when you have finished.

15. How do animals get the nitrogen that they need?

_____

_____

16. How is nitrogen returned to the atmosphere?

_____

_____

_____

_____

_____

Science and Engineering Practices

Think about how you have used models to show how cycles occur in an ecosystem. Tell how you can develop and use models by completing the "I can . . ." statement below.

Use examples from the lesson to explain what you can do!

I can _____

_____

_____

_____

Research, Investigate, and Communicate

Decomposers

 Explore the Digital Interactive *Decomposers* about the role of decomposers in ecosystems. Answer the questions when you have finished.

Crosscutting Concepts
Systems and System Models

1. List some decomposers and identify where they live.

_____

_____

_____

2. What does a decomposer do?

_____

_____

_____

3. Why are decomposers important to ecosystems?

_____

_____

_____

_____

_____

Performance Task
Plan a Terrarium

Think like a botanist and create an ecosystem for plants!

You will create a plan for building an ecosystem in the form of a terrarium.

Make a Model Consider what you need to model a terrarium in which all three cycles can successfully occur. List your materials and describe your plan below. Include a labeled sketch of what your finished terrarium will look like.

_____

_____

_____

_____

_____

Build a Terrarium

If time and materials allow, then build a terrarium based on your plan. Use the procedure and materials on this page as a guide. If time does not allow, then skip the procedure and answer the questions at the bottom of this page.

Make a Model

1. Place a thin layer of crushed charcoal in the bottom of the jar.

2. Place a few inches of topsoil over the charcoal.

3. Plant the plants in the soil.

4. Arrange the small rocks decoratively around the plants.

5. Add a small amount of water.

6. Place your terrarium in a well-lit spot, but avoid direct sunlight.

7. Maintain your terrarium by adding water once per week and occasionally poking holes in the soil with a pencil.

Materials

- ☐ wide-mouth jar or bottle
- ☐ soil
- ☐ small rocks
- ☐ crushed charcoal
- ☐ small plants such as moss, ferns, lucky bamboo
- ☐ water

Communicate Information

1. In what ways did your terrarium plan differ from the plan presented on this page?

_____

_____

_____

2. The plan above does not completely close the terrarium off from the outside environment. What do you think would happen if the terrarium above were sealed? Explain.

_____

_____

_____

?● Essential Question
How is matter cycled through ecosystems?

Think about the photo of a greenhouse you saw at the beginning of the lesson. Explain how matter cycles through an ecosystem.

_____

_____

_____

_____

_____

_____

_____

_____

⚙ Science and Engineering Practices

Review the "I can..." statement you wrote earlier in the lesson. Explain what you have accomplished in this lesson by completing the "I did..." statement.

I did... _____

_____

_____

Now that you're done with the lesson, share what you did!

Matter in Ecosystems

Performance Project
Composting At Home

Now it is your turn to think like an ecologist. Research different types of composting and composting activities. Create a composting plan or composting activity that you can try at home. Record your ideas below.

_____

_____

_____

_____

_____

Use your research to write a proposal for your school to implement a composting program. Draft your proposal here and write your final copy on a separate sheet of paper.

_____

_____

_____

_____

What can a compost pile help ecologists understand about ecosystems?

Draw a diagram of where the compost pile would be. Use labels to explain your diagram.

Explore More in Our World

Did you learn the answers to all of your questions from the beginning of the module? If not, how could you conduct research or design an experiment to help answer them?

Interactions of Earth's Major Systems

 ## Science in Our World

Ecological engineers design biodomes to study the interactions of living and nonliving things under certain conditions. What do you wonder about biomes and the role of ecological engineers?

_____

_____

_____

_____

_____

_____

abc Key Vocabulary

Look and listen for these words as you learn about Earth's systems.

atmosphere	biosphere	condensation
conservation	evaporation	geosphere
hydrosphere	precipitation	tectonic plates

> How do ecological engineers make the relationship between humans and Earth the best it can be?

MAYA
Geologist

STEM Career Connection

Ecological Engineer

As an ecological engineer, I am trying to find the best ways to integrate technology and ecosystems to benefit humans as well as living and nonliving things. I design and build biodomes, which are closed structures around natural ecosystems. We can use these structures to research ways to harness renewable sources of energy while keeping the ecosystems healthy.

How are ecological engineers working to increase renewable energy use?

_____

_____

_____

_____

_____

_____

_____

 # Science and Engineering Practices

I will obtain, evaluate, and communicate information.

I will develop and use models.

Name _____ Date _____

Earth's Major Systems

**PAGE KEELEY
SCIENCE
PROBES**

Earth's Systems

Four friends were talking about Earth's four systems: land, water, atmosphere, and life. They each had different ideas about what makes up Earth's four systems. This is what they said:

Tobias: *I think Earth's systems include the materials on Earth that interact.*

Jason: *I think Earth's systems include the processes on Earth that interact.*

Fay: *I think Earth's systems include the processes and materials on Earth that interact.*

Kristy: *I think Earth's systems include the processes and materials on Earth that interact as long as they are on land or in water.*

Who do you agree with most? _____

Explain why you agree

_____

_____

_____

Science in Our World

Look at the photo of Earth from space. What questions do you have about how the different types of matter on Earth interact with each other?

_____

_____

_____

_____

Read about a climate change analyst and answer the questions on the next page.

Climate change analysts gather information about the climate and communicate it to others.

STEM Career Connection
Climate Change Analyst

Over the past ten years, we have been tracking a variety of weather data in this region. We use computer models to analyze data from satellites that measure global temperatures, sea ice thickness, and rainfall distribution. This climate data will then be used to create mathematical models that help us predict changes to ocean and land temperatures in the next 50 to 100 years. We are hoping to present our findings at the next regional conference to make others aware of the patterns we have observed.

HIRO
Ocean Engineer

1. What does the climate change analyst do with the climate data from satellites?

_____

_____

_____

_____

2. What questions do you have about a climate change analyst's work?

_____

_____

_____

_____

? Essential Question

How do scientists define Earth's major systems?

_____

_____

_____

_____

_____

⚙ Science and Engineering Practices

I will obtain information to communicate.

> Like a climate change analyst, you will gather information and communicate it to others.

Inquiry Activity
Landform Models

How can you model Earth's surface to see differences in the types of landforms that exist?

Make a Prediction How can you map a model of landforms when you cannot see it? Explain.

_____

_____

_____

_____

Materials

- [ ] pencil
- [ ] clay
- [ ] shoe box with lid
- [ ] straw
- [ ] ruler
- [ ] graph paper

Carry Out an Investigation

BE CAREFUL Use caution poking holes in the lid.

1 Use the clay to make a model of a landscape in the bottom of the box.

2 Draw a 5 x 5 grid to fit the size of the lid. Label the squares A–E on the side and 1–5 across the top. See the tables on the next page to make sure they match.

3 Carefully poke a small hole at each corner on the grid. Put the lid on the box and trade models with another group.

4 Place the straw in each hole on the grid on the box lid. Measure how far the straw extends into each hole. Record your findings.

5 **Record Data** Use your results to create a graph showing the different landforms that are in the model.

Name _____ Date _____

Distance Straw Goes Into Box (cm)	1	2	3	4	5
A					
B					
C					
D					
E					

Color each box based on the elevation of the model in the shoe box. The greater the distance the straw went into the shoe box, the lower the elevation of that part of the model.

Black — 15 cm or more   Yellow — 3–5 cm
Purple — 12–14 cm   Orange — 1–2 cm
Blue — 9–11 cm   Red — less than 1 cm
Green — 6–8 cm

Distance Straw Goes Into Box (cm)	1	2	3	4	5
A					
B					
C					
D					
E					

Communicate Information

1. **Analyze Data** Take the lid off the box and compare your "map" to the model. Do the colors you used match the actual elevations in the model? How does this model compare to the one that you traded with another group?

_____

_____

_____

_____

_____

_____

2. What did you learn about Earth's systems by comparing the models in the activity?

_____

_____

_____

_____

_____

_____

_____

_____

Name _____   Date _____

 # Obtain and Communicate Information

abc Vocabulary

> **Use these words when explaining Earth's systems.**
>
> atmosphere   hydrosphere
>
> biosphere   geosphere

Record your predictions for the meanings of these vocabulary
word parts.

-sphere

_____

_____

atmo-

_____

_____

bio-

_____

_____

geo-

_____

_____

hydro-

_____

_____

FOLDABLES®

Cut out the Notebook Foldables given to you by your teacher.
Glue the anchor tabs as shown below. Use what you have learned
to describe each of the four spheres and their importance to life
on Earth.

Glue anchor tab here

Earth's Structures

📖 Read pages 138–139 in your *Science Handbook*. Use the definitions to check your predictions for the meanings of these vocabulary word parts. Record the correct definitions below.

-sphere

_____

atmo-

_____

bio-

_____

geo-

_____

hydro-

_____

How does understanding word parts help you figure out the meaning of unfamiliar words?

_____

_____

_____

⚙️ Science and Engineering Practices

Think about the information you have obtained about Earth's systems. Express how you can obtain information by completing the "I can..." statement below.

Use examples from the lesson to explain what you can do!

I can _____

_____

_____

⚙ Performance Task
Earth's Systems Poster

Materials

☐ poster board

☐ drawings or photos of objects in Earth's systems

☐ glue

☐ markers

You will create a poster that highlights how various parts of Earth's major systems interact with each other.

Make a Model Use information from this lesson and other appropriate sources to communicate information about how Earth's four major systems interact.

1. Choose specific features from each system to highlight those interactions.

2. Cut out or draw pictures of Earth's features and glue them to the poster board.

3. Use markers to draw arrows connecting different features and describe how those features are interacting.

4. Use the space below to plan your poster if you need to.

Communicate Information

⚙ Crosscutting Concepts
Systems and System Models

1. What features from each of Earth's four systems did you show on your poster?

_____

_____

_____

2. How did your poster model Earth's four systems interacting?

_____

_____

_____

_____

_____

_____

3. What other examples of interacting systems can you think of? Describe one below.

_____

_____

_____

_____

_____

_____

? Essential Question
How do scientists define Earth's major systems?

Think about the photo of Earth in space at the beginning of the lesson. How does that single image help scientists define Earth's major systems?

_____

_____

_____

_____

_____

_____

_____

⚙ Science and Engineering Practices

Review the "I can . . ." statement you wrote earlier in the lesson. Explain what you have accomplished in this lesson by completing the "I did . . ." statement.

Now that you're done with the lesson, share what you did!

I did _____

_____

_____

_____

Effects of the Geosphere

PAGE KEELEY SCIENCE PROBES

The Solid Earth

Three friends were talking about the solid parts of Earth. They each had different ideas about where Earth's rock, soil, and sediments are found. This is what they said:

Francis: *I think the solid materials on Earth are found on land.*

Portia: *I think the solid materials on Earth are found under the oceans.*

Trent: *I think the solid materials on Earth are found on land and under the oceans.*

Who do you agree with the most? _____

Explain why you agree.

_____

_____

_____

_____

Science in Our World

Look at the map of Pangaea. How is it different from modern maps of the world? What questions do you have about this map?

_____

_____

_____

_____

_____

Read about a soil scientist and answer the questions on the next page.

Soil scientists study Earth's soils and how they are formed.

STEM Career Connection
Soil Scientist

I am a soil scientist for the United States Geologic Survey, or USGS. I research how well different types of soils hold together or fall apart in different environmental conditions. Different kinds of soil are made of different minerals. Some soils are moist and held together by plant roots. Other soils are dry and loose and can be blown away with the wind.

I study the behavior of these soils to understand how they interact with wind and water. Predicting the behavior of Earth's materials is an important task. It can protect people and property from natural hazards such as earthquakes, floods, and landslides.

HIRO
Ocean Engineer

1. Why is it important to understand the behavior of Earth's materials, such as soil?

_____

_____

_____

_____

2. What questions do you have about a soil scientist's work?

_____

_____

_____

? Essential Question
How does the geosphere affect other systems?

_____

_____

_____

_____

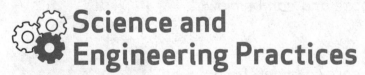 ## Science and Engineering Practices

I will develop and use models.

Like a soil scientist, you will create and use models to show how the geosphere affects other systems.

Inquiry Activity
Modeling Earth Movements

How can a model allow you to observe the movement of Earth's crust during an earthquake?

Make a Prediction What might happen to roads and buildings during earthquakes?

_____

_____

Materials

☐ tape

☐ construction paper

☐ two textbooks

☐ marker

☐ building blocks or dominoes

Carry Out an Investigation

1. Tape two pieces of construction paper to the covers of the similar-sized textbooks.

2. Place the books next to each other. Draw a "road" that crosses from one piece of construction paper to the other.

3. Use building blocks or dominoes to construct a small "house" where the books meet.

4. Slowly slide one book past the other. Observe what happens to the road and the house.

5. **Record Data** Record your observations by drawing a sketch and labeling the details of what you observed.

6 Line up the road and rebuild your house to match its appearance in step 3.

7 Tap one book sharply along its shorter side. Observe what happens to the road and the house.

8 **Record Data** Record your observations by drawing a sketch and labeling the details of what happened.

Communicate Information

1. What happened to the blocks and road when you started to move the books apart? Why do you think this happened?

_____

_____

_____

_____

2. **Construct an Explanation** How do you think this activity modeled movements on Earth's surface?

_____

_____

_____

_____

Obtain and Communicate Information

🔤 Vocabulary

> **Use these words when explaining changes to Earth's surface.**
>
> crust   mantle   core
>
> continental drift   tectonic plates   fault

Earth's Layers

📖 Read pages 140–141 in the *Science Handbook*. Answer
   the following questions after you have finished reading.

1. Name and describe the three main layers of the geosphere.

_____

_____

_____

_____

_____

_____

2. Look at the diagram on page 141. Match the numbers on the map
   of the United States with the type of landform. What landform and
   number is in each region?

_____

_____

_____

_____

Name _____ Date _____

FOLDABLES®

Cut out the Notebook Foldables given to you by your teacher.
Glue the anchor tabs as shown below. Use what you have learned
and the *Science Handbook* to define the terms.

Glue anchor tab here.

Glue anchor tab here

Plate Tectonics

📖 Read pages 172–177 in the *Science Handbook.* Answer the following questions after you have finished reading.

3. Describe Alfred Wegener's theory and the evidence that supports it.

_____

_____

_____

_____

_____

_____

4. Explain the theory of plate tectonics.

_____

_____

_____

_____

_____

_____

5. What causes folded mountains?

_____

_____

_____

_____

Earthquakes

Explore the Digital Interactive *Earthquakes* that describes the effects of earthquakes on the biosphere. Answer the following question when you have finished.

6. What is one possible effect of an earthquake?

_____

Earth's Soils

Read pages 158–161 in the *Science Handbook*. Answer the following questions after you have finished reading.

7. What is soil made of?

_____

_____

_____

8. How does soil form?

_____

_____

_____

_____

9. How are the three types of soils different?

_____

_____

_____

_____

Mountain Ranges

📖 Read page 212 in the *Science Handbook.* Answer the following question after you have finished reading.

10. Describe how a rain shadow forms.

_____

_____

_____

_____

_____

_____

🔊 Explore the Digital Interactive *Water on Earth* on how water pools in depressions on Earth's surface. Answer the following question after you have finished.

11. What causes water to move on Earth's surface?

_____

_____

⚙️ # Science and Engineering Practices

Think about your landform model and what you have learned about the geosphere. Tell how you can use models by completing the "I can . . ." statement below.

I can _____

_____

_____

> Use examples from the lesson to explain what you can do!

Name _____ Date _____

Inquiry Activity
Landform Models Part 2

How can you use a model to show how the geosphere, atmosphere, and hydrosphere interact?

Make a Prediction What will happen when water is sprinkled on the model you made in Lesson 1?

_____

_____

_____

Materials
☐ safety goggles
☐ plastic tray or container
☐ landform models from Lesson 1
☐ watering can
☐ water

Make a Model

BE CAREFUL Wear safety goggles. Use caution to avoid spills.

1 Sprinkle a small amount of water uniformly on your model to simulate rain over a land surface.

2 Observe what happens to the water and the land surface.

3 **Record Data** Record you observations below.

_____

_____

_____

_____

_____

Communicate Information

1. Which of Earth's systems were modeled in this activity?

_____

_____

2. What happened to the land and water in this model?

_____

_____

_____

3. How did the land affect where the water went?

_____

_____

_____

4. **Construct an Explanation** How do models help scientists gather information and understand Earth's systems?

_____

_____

_____

_____

_____

⚙️ Performance Task
Rain Shadow

You will build a model of a mountain range to demonstrate how mountains affect the weather that occurs around them.

Make a Model

Materials

☐ modeling clay

☐ cardboard squares for base

☐ acrylic or tempura paint

☐ paintbrush

☐ cotton balls

☐ glue

1. Place a mound of clay on the base and begin to shape it into a cone-shaped mountain. Add more clay to make one large mountain or make several smaller mountains to form a mountain range.

2. Use your fingers to make finer details in the clay. Create ridges along the sides of the mountain(s) and define the valleys if you are making a mountain range. Make the sides of the mountains vary between steep sides and gradual slopes. Many mountains are very steep near the top, and then they slope gradually near the bottom.

3. Designate a tree line on the mountains by making a rocky area up near the top of the mountain and smoother sides at the bottom.

4. Paint your model to indicate grass, trees, bare rock, soil, snow, and water features. Allow it to dry. Glue several cotton balls together to use as a cloud in your model.

Think like a soil scientist and show how mountains affect the weather.

Communicate Information

1. **Construct an Explanation** Explain how a rain shadow forms.
   Write your explanation so that you can read it while demonstrating
   with your model.

_____

_____

_____

_____

_____

_____

_____

_____

_____

_____

_____

_____

_____

_____

⚙ Crosscutting Concepts
Systems and System Models

2. Describe which of Earth's systems were represented in your model.

_____

_____

_____

_____

_____

_____

3. Describe how these systems interacted in your model and demonstration.

_____

_____

_____

_____

_____

_____

4. How would a soil scientist characterize the differences in the soils on opposite sides of the mountain?

_____

_____

_____

_____

_____

_____

_____

❓ Essential Question
How does the geosphere affect other systems?

Think about the map of Earth's continents you saw at the beginning of the lesson. Use what you have learned to explain how the same forces that can move continents in the geosphere affect Earth's other systems.

_____

_____

_____

_____

_____

_____

⚙️ Science and Engineering Practices

Review the "I can . . ." statement you wrote earlier in the lesson. Explain what you have accomplished in this lesson by completing the "I did . . ." statement.

Now that you're done with the lesson, share what you did!

I did _____

_____

_____

_____

_____

Effects of the Hydrosphere

What Covers Earth?

Three friends wondered what covered most of Earth's surface.
They each had different ideas. This is what they said:

Mia: *I think Earth is covered mostly by ice and snow.*

Sam: *I think Earth is covered mostly by water.*

Cate: *I think Earth is covered mostly by land.*

Who do you agree with the most? _____

Explain why you agree.

_____

_____

_____

_____

Science in Our World

▶ Watch the video of waves crashing onto the shore.
What questions do you have about waves?

_____

_____

_____

_____

_____

Read about a desalination engineer and answer
the questions on the next page.

Desalination engineers
find ways to take the
salt out of ocean water.

STEM Career Connection
Desalination Engineer

In my job as a desalination engineer, I have important
duties at a water treatment plant. I remove salt and other
chemicals from water to make it safe for people to drink and
use. Most of the water on Earth is salt water. Only a small
percentage is fresh water, and many freshwater sources
are not clean enough to drink.

In nature, the energy of the Sun evaporates sea
water and leaves the salt behind. Many desalination
plants use a lot of energy to complete this same task,
but new technologies with special filters make the
process more efficient. This is important because some
places in the world do not have enough clean water to
meet people's needs.

HIRO
Ocean Engineer

1. What does a desalination engineer do to make water safe for people to use?

_____

_____

_____

_____

_____

2. What ways do you use water every day?

_____

_____

_____

? Essential Question

How does the hydrosphere affect other systems?

_____

_____

_____

⚙ Science and Engineering Practices

I will obtain, evaluate, and communicate information.

Like a desalination engineer, you will evaluate information and communicate it to others.

Inquiry Activity
Using Water

How much water do you usually use? How can knowing how much water we use help us conserve water?

Make a Prediction How many gallons of water do you use every day washing your hands?

_____

_____

_____

Carry Out an Investigation

BE CAREFUL Wear safety goggles. Use caution to avoid spills.

1 Place a large container under the faucet in the sink.

2 Turn on the water and pretend to wash your hands (without using soap) for the same amount of time that you normally would.

3 Use the measuring cup to carefully scoop the water out of the container and into the sink. Keep track of how many cups of water you scoop.

4 **Record Data** Number of cups of water used: _____ cups

5 **Analyze Data** Calculate how many cups of water you use in one day washing your hands if you wash your hands 10 times per day. Number of cups of water used: _____ × 10 times per day = _____ cups per day

6 There are 16 cups in one gallon. About how many gallons of water do you use each day washing your hands? Number of cups of water used in one day: _____ ÷ 16 cups per gallon = about _____ gallons

Materials

- [ ] safety goggles
- [ ] large plastic container
- [ ] sink
- [ ] measuring cup
- [ ] calculator
- [ ] graph paper

7 Calculate how many cups of water you use washing your hands in one week if you wash your hands 10 times per day. Number of cups of water used in one day: _____ × 7 days per week = _____ cups

8 About how many gallons of water do you use to wash your hands in one week? Number of cups of water used in one week: _____ ÷ 16 cups per gallon = about _____ gallons

9 Calculate how many cups of water you use to wash your hands in one year if you wash your hands 10 times per day. Number of cups of water used in one day: _____ × 365 days per year = _____ cups

10 About how many gallons of water do you use to wash your hands each year? Number of cups of water used in one year: _____ ÷ 16 cups per gallon = about _____ gallons

Crosscutting Concepts
Scale, Proportion, and Quantity

Communicate Information

1. How would setting a timer affect how much water you used to wash your hands?

_____

_____

2. What are some other activities that you could test to see how much water you use?

_____

_____

3. Besides setting a timer, how could you reduce the amount of water you use washing your hands?

_____

_____

Obtain and Communicate Information

🔤 Vocabulary

> **Use these words when explaining how the hydrosphere affects Earth's other systems.**
>
> groundwater   aquifer   estuary
>
> erosion   deposition   desalination

Earth's Water

📖 Read pages 154–156 in the *Science Handbook*. Answer the following questions after you have finished reading.

1. What percentage of Earth's surface water is salt water?

_____

2. Describe where Earth's fresh water is found and how much is in each form.

_____

_____

_____

_____

3. What is an aquifer?

_____

_____

_____

4. What are the two most abundant minerals in ocean water?

_____

_____

Water Ecosystems

📖 Read pages 94–96 in the *Science Handbook*. Answer the following questions after you have finished reading.

5. What abiotic factors determine which organisms can survive in a water ecosystem?

_____

_____

_____

6. Why do few organisms live in the deepwater bottom zone of a standing water ecosystem?

_____

_____

7. What is an estuary?

_____

_____

_____

_____

8. Describe the four ocean ecosystem zones.

_____

_____

_____

_____

_____

_____

_____

FOLDABLES®

Cut out the Notebook Foldables given to you by your teacher.
Glue the anchor tabs as shown below. Use what you have learned
to describe the effects of water being added to or removed from
a habitat.

Glue anchor tab here

Hydrosphere Effects

📖 Read pages 209 and 166–171 in the *Science Handbook*.
Answer the following questions after you have finished reading.

9. How does distance from a body of water affect the weather on land?

_____
_____

10. How does running water cause erosion?

_____
_____

11. How does a glacier cause erosion?

_____
_____

12. How do waves affect the shoreline?

_____
_____
_____

⚙ Science and Engineering Practices

Use examples from the lesson to explain what you can do!

Think about the information you have obtained about Earth's hydrosphere. Tell how you can obtain, evaluate, and communicate information about the water on Earth by completing the "I can . . ." statement below.

I can _____
_____

Research, Investigate, and Communicate

 Inquiry Activity
Clean Water

Many people in the world do not have easy access to clean, fresh water.

Make a Prediction Can you tell whether water is clean just by looking at it?

_____

_____

Materials

☐ safety goggles

☐ clear cup or glass

☐ water

☐ teaspoon

☐ vinegar

☐ salt

☐ sugar

Carry Out an Investigation

BE CAREFUL Wear safety goggles. Use caution to avoid spills.

1 Fill the glass two-thirds full of water.

2 Add one teaspoon of vinegar to the water and stir slowly.

3 Add one teaspoon of salt to the water and stir slowly.

4 Add one teaspoon of sugar to the water and stir slowly.

5 **Record Data** Use your observations to answer the following questions.

1. Did the appearance of the water change when you added each material?

_____

_____

_____

2. What does this demonstrate about the appearance of water?

_____

_____

_____

Research, Investigate, and Communicate

Ways Water is Used

▶ Watch *Ways Water Is Used* about how water is used around the world. Answer the questions after you have finished watching.

3. Compare the ways you use water with the examples in the video.

_____

_____

_____

_____

_____

_____

4. Would you be able to do the tasks on your list if you did not have clean water? List whether each task could or could not be done without clean water.

_____

_____

_____

_____

_____

_____

_____

_____

_____

Water Treatment

👁 Read *Water Treatment* on ways water is cleaned for human use. Answer the following questions after you have finished reading.

5. List and describe the steps followed in a water treatment plant.

_____

_____

_____

_____

_____

6. How can evaporation be used to clean water? Would it be able to separate harmful substances? Explain.

_____

_____

_____

_____

_____

_____

7. Why is a clean environment important for our drinking water?

_____

_____

_____

_____

_____

_____

Performance Task
Where Water Is Found

Materials
- [ ] world map
- [ ] graph paper
- [ ] colored pencils

Earth is covered with water. The two types of water are fresh water and salt water. Earth's water features come in many shapes and sizes.

Make a Model Obtain and evaluate information about the types of water on Earth's surface. Use the information to make a model of the amounts of salt water and fresh water on Earth.

1. Look at a world map in your classroom or online.

2. **Record Data** Make a list of all the types of water features you can see on the map. Use your *Science Handbook* to help you.

_____

_____

_____

_____

Think like a desalination engineer and show how much water is available for human use.

3. **Analyze Data** Rank the water features you found on the map from largest to smallest.

_____

_____

_____

4. Classify each water feature as either fresh water of salt water. Record them in the table below.

Salt Water	Fresh Water

You will create a graph that shows the amount of salt water, frozen fresh water, and available fresh water. Use pages 154–155 in the *Science Handbook* to find the data you need.

5 On a sheet of graph paper, draw a 10 x 10 square. Across the top of the page, add the title *Water on Earth*.

6 Within this square, count out the number of individual squares equal to the percent of salt water on Earth. Color these squares green. Label this area *Salt Water*.

7 Count the number of individual squares equal to the percent of fresh water in glaciers to the right of the green box. Color these squares yellow. Label this area *Glaciers*.

8 Count the number of individual squares equal to the percent of other sources of fresh water to the right of the white box. Color these squares blue. Label this area *Groundwater, Lakes, and Rivers*.

Communicate Information

1. **Construct an Explanation** What does the graph tell you about water on our planet?

_____

_____

_____

2. What does the graph suggest about the importance of desalination?

_____

_____

_____

_____

Glue your graph here.

Design a Solution

Think about what you have learned about the amount of clean water on Earth. Design an original solution that will give people without easy access to clean water the water they need. Write about and sketch your solution below.

_____

_____

_____

_____

_____

_____

_____

_____

? Essential Question
How does the hydrosphere affect other systems?

▶ Think about the video of waves crashing on the shore from the beginning of the lesson. Use what you have learned to explain how the hydrosphere affects other Earth systems.

_____

_____

_____

_____

_____

_____

_____

⚙ Science and Engineering Practices

Review the "I can . . ." statement you wrote earlier in the lesson. Explain what you have accomplished in this lesson by completing the "I did . . ." statement.

I did _____

_____

_____

_____

Now that you're done with the lesson, share what you did!

Name _____ Date _____

Effects of the Atmosphere

**PAGE KEELEY
SCIENCE
PROBES**

What Happened to the Puddle?

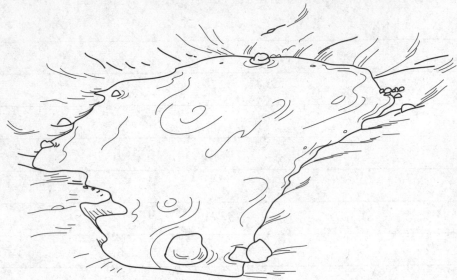

There were puddles on the sidewalk in the morning when Kevin and his friends walked to school. When they walked home, the puddles were gone. They each had different ideas about where the water that was in the puddles went. This is what they said:

Kevin: *I think it went up to the Sun.*

Lucy: *I think it went up to the clouds.*

Doug: *I think it went into the air.*

Becca: *I think it just disappeared.*

Who do you agree with the most? _____

Explain why you agree.

_____

_____

_____

_____

Science in Our World

Look at the photos of the cloud formations. What types of weather do you think will happen with these clouds? What questions do you have?

_____

_____

_____

_____

_____

Read about a meteorologist and answer the questions on the next page.

Meteorologists gather information and predict weather patterns.

STEM Career Connection
Meteorologist

As a meteorologist at the National Oceanic and Atmospheric Administration, or NOAA, I use different kinds of instruments to collect data about the variables that can affect a storm. I obtain information from weather stations around the world. These stations use instruments such as weather vanes, barometers, and rain gauges to gather data about local weather conditions.

Satellite photos help meteorologists predict upcoming weather. We can forecast what the weather will be for the next day, five days, or even the next month. However, our weather predictions might not always be accurate. I constantly have to measure the variables that affect weather because they can change very quickly.

HIRO
Ocean Engineer

1. How does a meteorologist gather information to predict weather?

_____

_____

_____

_____

2. Why aren't weather predictions always accurate?

_____

_____

_____

_____

? Essential Question
How does the atmosphere affect other systems?

_____

_____

_____

_____

⚙ Science and Engineering Practices

I will develop and use models.

Like a meteorologist, you will use models to analyze the effects of the atmosphere.

Inquiry Activity
Warm and Cold Air Masses

How can the interactions of air masses help us predict what type of weather will occur?

Make a Prediction What happens when air masses of different temperatures meet?

_____

_____

Carry Out an Investigation

BE CAREFUL Use caution to avoid spills and when using scissors to cut the cardboard.

1 Use scissors to cut the cardboard so it fits tightly in the clear box. Wrap the cardboard in aluminum foil.

2 Hold the cardboard tightly against the bottom of the box. Pour four cups of cold water into one side of the box and four cups of warm water into the other side of the box.

3 Place a few drops of blue food coloring into the cold water and a few drops of red food coloring into the warm water.

4 Watch the box from the side as you remove the cardboard.

5 **Record Data** Record your observations below.

_____

_____

_____

_____

6 **Analyze Data** What can you infer about what happens when warm and cold air masses meet in Earth's atmosphere?

_____

_____

Materials

- [ ] scissors
- [ ] cardboard
- [ ] clear plastic box
- [ ] aluminum foil
- [ ] 4 cups of cold water
- [ ] measuring cup
- [ ] 4 cups of warm water
- [ ] red and blue food coloring

7 Now repeat the same test with water of similar temperature on both sides and food coloring only in one side.

8 **Record Data** Record your observations below.

_____

_____

9 **Analyze Data** What can you infer from this about what happens when two warm air masses meet in Earth's atmosphere?

_____

_____

Communicate Information

1. You learned in Lesson 2 that when a mountain blocks an air mass, the air mass is forced to rise. Rising air cools and loses its ability to hold moisture, causing precipitation. Explain whether a cold air mass meeting a warm air mass is a similar or different situation.

_____

_____

_____

_____

2. Would a cold air mass meeting a warm air mass cause precipitation? Explain.

_____

_____

_____

_____

_____

Obtain and Communicate Information

abc Vocabulary

> **Use these words when explaining how the atmosphere affects Earth's other systems.**
>
> evaporation   condensation   precipitation
>
> climate   front

Weather and Climate

📖 Read pages 186–189, 192–195, and 208 in the *Science Handbook.* Answer the following questions after you have finished reading.

1. What are the four layers of Earth's atmosphere?

_____

_____

_____

2. What happens to the Sun's energy when it reaches Earth?

_____

_____

_____

3. Define *weather.*

_____

_____

4. Define *climate.*

_____

_____

The Water Cycle

▦ Investigate how water cycles on Earth by conducting the *simulation*. Answer the questions after you have finished.

5. What kind of energy causes liquid water to transform into a gas?

_____

6. How are clouds formed?

_____

_____

_____

7. How do clouds move from one place to another?

_____

_____

8. Name two ways that water is stored on Earth.

_____

_____

_____

Causes of Evaporation

🔊 Explore the Digital Interactive *Causes of Evaporation* about what fuels the water cycle. Answer the questions after you have finished.

9. Why is heat energy needed for the water cycle?

_____

_____

_____

10. What is evaporation?

_____

_____

FOLDABLES®

Cut out the Notebook Foldables given to you by your teacher.
Glue the anchor tabs as shown below. Use what you have learned to
describe how each stage of the water cycle might be affected by
the seasonal changes in the atmosphere.

Glue anchor tab here

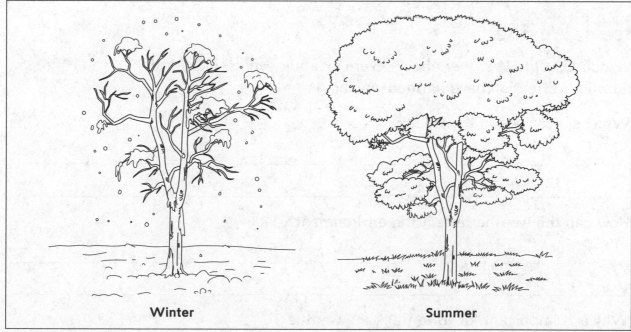

Winter   Summer

Climate and Plants

📖 Read page 213 in the *Science Handbook.* Answer the following questions after you have finished reading.

11. Explain how an area's climate can be described by the plants that grow there.

_____

_____

_____

12. Which atmospheric gases absorb radiated heat energy?

_____

_____

13. How does burning fossil fuels affect Earth's atmosphere?

_____

_____

_____

Extreme Weather

▶ Watch *Extreme Weather* about severe weather events. Answer the questions after you have finished watching.

14. What did you see in the video?

_____

_____

15. How can the weather affect the environment?

_____

_____

16. Why is it important for us to track the weather?

_____

_____

Physical Weathering

📖 Read pages 162–163, in the *Science Handbook*.
Answer the following question after you have finished reading.

17. Explain how events connected with the atmosphere can weather rocks.

_____

_____

_____

_____

_____

⚙️ Science and Engineering Practices

Think about how you have developed and used models of the atmosphere. Tell how you can use these models to explain how the atmosphere affects other systems by completing the "I can . . ." statement below.

Use examples from the lesson to explain what you can do!

I can _____

_____

_____

_____

_____

_____

Weather

📖 Read pages 197–201, in the *Science Handbook*. Answer
the following questions after you have finished reading.

1. What are the four different types of air masses, and where
do they originate?

_____

_____

_____

_____

_____

2. Describe weather that occurs with the three different types of fronts.

_____

_____

_____

_____

3. Describe some of the tools that meteorologists use and what types of
data they collect.

_____

_____

_____

_____

_____

_____

4. Explain how the weather differs between a high-pressure system
and a low-pressure system.

_____

_____

_____

⚙ Performance Task
Water Cycle Model

You will create a model of the water cycle. Follow the directions below or create your own plan.

▦ Before building your model, revisit the *Water Cycle* simulation to review each of the stages.

<table>
<tr><td>Materials</td></tr>
<tr><td>☐ small, clear plastic cup</td></tr>
<tr><td>☐ marker</td></tr>
<tr><td>☐ water</td></tr>
<tr><td>☐ large resealable bag</td></tr>
<tr><td>☐ tape</td></tr>
</table>

Make a Model

BE CAREFUL Use caution to avoid spills.

1 Fill the plastic cup halfway with water. Draw a line at the top of the water with the marker.

2 Open the resealable bag and hold it by one of the top corners. Carefully stand the cup of water up in the opposite bottom corner of the resealable bag.

3 Seal the bag, leaving some air inside. Make sure you hold the bag upright so you do not spill the water that is in the cup.

4 Carefully tape the bag against a window that receives sunlight.

5 Over the next day or two, make observations of the water cycle model you have made.

6 Draw a diagram of your model in the space below. Label the following processes on your model: evaporation, condensation, precipitation, runoff, and storage.

Communicate Information

1. **Construct an Explanation** Use your model to explain how water flows through the water cycle to a partner. Mention at each step whether the water is part of the hydrosphere or part of the atmosphere. Write your explanation so you can read it while you are demonstrating with your model.

_____

_____

_____

_____

_____

_____

_____

Crosscutting Concepts
Systems and System Models

2. Does the water cycle have a beginning or end? Explain.

_____

_____

_____

3. Would the water cycle exist without evaporation? Explain.

_____

_____

_____

_____

? Essential Question
How does the atmosphere affect other systems?

Think about the photos of the cloud formations from the beginning of the lesson. Use what you have learned to describe how the atmosphere affects Earth's other systems.

_____

_____

_____

_____

_____

⚙ Science and Engineering Practices

Review the "I can . . ." statement you wrote earlier in the lesson. Explain what you have accomplished in this lesson by completing the "I did . . ." statement.

Now that you're done with the lesson, share what you did!

I did _____

_____

_____

_____

Effects of the Biosphere

PAGE KEELEY
SCIENCE
PROBES

Use of Resources

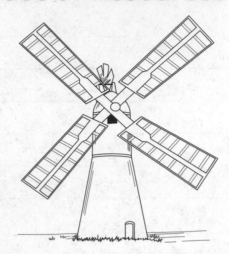

Four friends were talking about how humans use renewable and nonrenewable resources. They each had different ideas. This is what they said:

Wyatt: *When humans use renewable resources, they can harm the environment. Nonrenewable resources do not harm the environment.*

Sarah: *When humans use renewable resources, they do not harm the environment. Nonrenewable resources can harm the environment.*

Hector: *Both renewable and nonrenewable resources can harm the environment when humans use them. Just because some resources are renewable doesn't mean they have no harmful effects.*

Gus: *Renewable and nonrenewable resources do not harm the environment if they are natural resources. Natural resources are not harmful to the environment.*

Who do you agree with most? _____

Explain why you agree.

_____

_____

_____

Science in Our World

Look at the photos of solar panels. What are they used for?
What questions do you have about this type of energy?

_____

_____

_____

_____

_____

Read about an energy efficiency engineer and answer
the questions on the next page.

> An energy efficiency engineer helps us use less energy to power our machines.

STEM Career Connection
Energy Efficiency Engineer

How much energy do you think all of the houses
and buildings in the country use? As an energy efficiency
engineer, I have found that they use up to 40% of all
the energy that the United States produces in a year!
The number is high because all those houses and buildings
need to be heated, cooled, ventilated, lit, and supplied
with hot water and electricity to run appliances and
computers.

Energy efficiency engineers help reduce the
amount of energy houses and buildings use.
I recommend installation of the most efficient
lighting and heating systems. I review architectural
plans before a building is built to point out areas
that waste energy. I also conduct an energy
audit using an infrared camera to spot energy
loss in a building. This saves families and businesses
money and lowers the amount of energy used.

HIRO
Ocean Engineer

1. What is the main job of an energy efficiency engineer?

_____

_____

_____

_____

_____

2. What questions do you have about energy efficiency?

_____

_____

_____

_____

? Essential Question
How does the biosphere affect other systems?

_____

_____

_____

⚙ Science and Engineering Practices

I will obtain, evaluate, and communicate information.

Like an energy efficiency engineer, you will obtain, evaluate, and communicate information about human impact on the environment.

🖐 Inquiry Activity
Mining Cookies

How does mining affect Earth's surface and resources?

Make a Prediction Can you remove a chocolate chip from a cookie without damaging the cookie?

_____

_____

_____

Carry Out an Investigation

BE CAREFUL Do not eat lab materials. Use caution with sharp objects.

1 Place one cookie on a plate.

2 Using a toothpick, try to remove all of the chocolate chips you can see without causing any cracks or breaks in the cookie.

3 **Analyze Data** After removing all of the chips you can see, examine the cookie carefully. Answer the following questions.

1. How does removing the chips change the surface of the cookie?

_____

_____

_____

2. Can you make the cookie look the way it did before you started?

_____

_____

_____

3. How is this model similar to mining on Earth?

_____

_____

_____

Research Gather information from print and online sources about how coal, iron, copper, and other minerals are mined from beneath Earth's surface. Read the following questions before you begin so you know what information to look for.

Communicate Information

4. How are coal and other ores mined?

_____

_____

_____

_____

5. What are some positive outcomes of mining?

_____

_____

_____

6. What are some negative consequences of mining?

_____

_____

7. How could a mining area be improved to make it more attractive or useful?

_____

_____

_____

Obtain and Communicate Information

🔤 Vocabulary

Use these words when explaining how the biosphere affects Earth's other systems.

nonrenewable resources   renewable resources

pollution   conservation

Living Things in Earth's Systems

🔊 Explore the Digital Interactive *Living Things in Earth's Systems* about living things using Earth's other systems. Answer the question after you have finished.

1. How do plants affect Earth's other systems?

_____

_____

_____

Humans and the Environment

📖 Read pages 126–129 in the *Science Handbook*. Answer the following questions after you have finished reading.

2. What is a natural resource? List some examples.

_____

_____

_____

3. What is a nonrenewable resource? List some examples.

_____

_____

_____

4. What is a renewable resource? List some examples.

_____

_____

_____

_____

Pollution in Earth's Systems

👁 Read *Pollution in Earth's Systems* on the effects of some human activities in Earth's systems. Answer the following questions after you have finished reading.

5. In what ways are the oceans being polluted?

_____

_____

_____

6. How can humans make a positive choice to impact Earth's systems?

_____

_____

_____

7. Construct an Explanation What choices can humans make to help the environment?

_____

_____

_____

_____

8. What can you personally do to help the environment?

_____

_____

FOLDABLES®

Cut out the Notebook Foldables given to you by your teacher. Glue the anchor tabs as shown below. Use what you have learned to explain how humans impact the environment when they use these resources.

Glue anchor tab here

Conservation

📖 Read pages 131–135 in the *Science Handbook.* Answer the following questions after you have finished reading.

9. What is conservation?

_____

_____

10. What are three ways the U.S. Government has worked to protect the environment?

_____

_____

_____

11. What are some threats to ecosystems?

_____

_____

_____

⚙️ Science and Engineering Practices

Think about the information you have obtained about the biosphere and its effects on Earth's other systems. Tell what you can do with that information by completing the "I can . . ." statement below.

> Use examples from the lesson to explain what you can do!

I can _____

_____

_____

Research, Investigate, and Communicate
Effects of Acid Rain

🔊 Explore the Digital Interactive *Effects of Acid Rain* on the cause and effect of polluted rain water. Answer the question after you have finished.

1. How does rain become acid rain? What does acid rain do?

_____

_____

_____

✋ Inquiry Activity
Effects of Acid Rain

You will investigate how acid rain can cause erosion on rocks.

Make a Prediction How long does it take acid rain to dissolve rock material?

_____

_____

Materials
☐ safety goggles
☐ paper plate
☐ small cup of vinegar solution
☐ chalk
☐ dropper

Carry Out an Investigation

BE CAREFUL Wear safety goggles. Do not inhale vinegar fumes.

1️⃣ Place a piece of chalk on a paper plate.

2️⃣ Using the dropper, carefully place two drops of the vinegar solution directly onto the chalk.

3️⃣ Observe the reaction. If necessary, repeat step 2.

Communicate Information

2. **Record Data** What happened when the acid touched the chalk?

_____

_____

3. **Analyze Data** Is it possible to make the chalk return to its original state? Explain.

_____

_____

_____

Crosscutting Concepts
Science Addresses Questions About the Natural and Material World

4. Why is acid rain partly a result of human choices?

_____

_____

5. **Construct an Explanation** Does acid rain damage other materials besides stone?

_____

_____

_____

6. What can be done to minimize or prevent further acid rain damage?

_____

_____

_____

Performance Task
Human Impact Research Project

Research Choose one way humans impact the environment and research it thoroughly. Define the activity as having a positive or negative effect on the environment. Research community groups that support or are critical of the activity. Record your notes below.

_____

_____

_____

_____

_____

_____

_____

_____

_____

_____

_____

_____

Think like an energy efficiency engineer and look for ways we can improve our world.

Communicate Information

1. **Construct an Explanation** How would you create a campaign to support the positive impact of the activity or reduce the negative human impact of the activity?

_____

_____

_____

_____

2. Organize your information on a poster to present to the class. Create a draft of your poster below.

? Essential Question
How does the biosphere affect other systems?

Think about the solar panels you saw at the beginning of the lesson. What are some negative and positive ways that humans are affecting Earth's systems?

_____

_____

_____

_____

_____

_____

_____

Science and Engineering Practices

Review the "I can . . ." statement you wrote earlier in the lesson. Explain what you have accomplished in this lesson by completing the "I did . . ." statement.

Now that you're done with the lesson, share what you did!

I did _____

_____

_____

_____

Interactions of Earth's Major Systems

⚙ Performance Project
Plan a Planet

You will use what you have learned about the interactions of Earth's major systems to plan a model of a habitat on another planet. In your plan, name and describe your planet and include ways that the geosphere, biosphere, hydrosphere, and atmosphere interact. Write your plan below.

_____

_____

_____

_____

_____

_____

_____

_____

_____

_____

_____

_____

_____

_____

_____

How do ecological engineers make the relationship between humans and Earth the best it can be?

Make a diagram to model your planet below showing at least one of the major systems in action. Label your diagram. Share your ideas and thoughts with your classmates.

If time and materials allow, make a three-dimensional model of your planet.

 Explore More in Our World

Did you learn the answers to all of your questions from the beginning of the module? If not, how could you design an experiment or conduct research to help answer them?

The Solar System and Beyond

 ## Science in Our World

Objects in space have some incredible and unique qualities. Look at the photos of our solar system. Why do you think it is organized this way? How are the space objects alike and different?

_____

_____

_____

_____

_____

_____

_____

_____

🔤 Key Vocabulary

Look and listen for these words as you learn about the solar system.

asteroid	comet	constellation
gravity	inertia	meteor
orbit	phase	revolution
rotation	star	tide

How can a space scientist analyze data to show how space objects interact?

HALEY
Astronomer

STEM Career Connection
Space Scientist

Our latest observations show that sunspot levels are increasing, but continue to be far below average. This indicates that the Sun has entered a period of unusual low magnetic activity. It's amazing to think that these tiny dark spots we see on the surface of the Sun are actually brighter than the full Moon, and many are much larger than Earth. We know that sunspots are caused by solar magnetic fields in the outer layer of the Sun, but we don't know what causes the number of sunspots to change. Some scientists have proposed that there seems to be a connection between sunspot activity and weather patterns on Earth. This connection does not imply cause and effect.

Why is it important for space scientists to continue observing the Sun?

_____

_____

_____

_____

Science and Engineering Practices

I will engage in argument from evidence.

I will analyze and interpret data.

Movements of the Sun, Earth, and the Moon

 **PAGE KEELEY SCIENCE PROBES**

Earth and the Sun

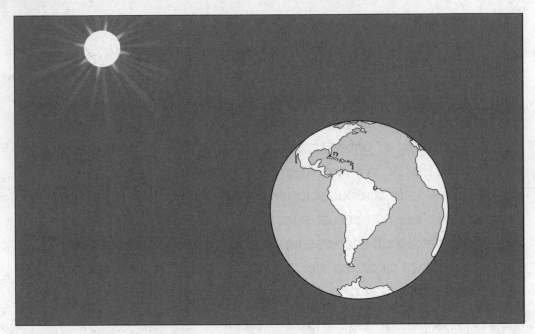

Emmy and Dexter were talking about Earth and the Sun. They each had different ideas about movement in Earth-Sun system. This is what they said:

Emmy: *I think the Earth moves around the Sun.*

Dexter: *I think the Sun moves around Earth.*

Who do you agree with most? _____

Explain why you agree.

_____

_____

_____

_____

Science in Our World

Look at the photos of the polar night and the midnight Sun. What questions do you have about how the Sun, Earth, and the Moon interact?

_____

_____

_____

_____

_____

Read about an aerospace engineer and answer the questions on the next page.

> Aerospace engineers analyze and interpret data about other planets.

STEM Career Connection
Aerospace Engineer

I design rovers as an aerospace engineer for the National Aeronautics and Space Administration, also known as NASA. Rovers are small, wheeled vehicles that are transported to other planets aboard spacecraft. Once they are on the surface of a new planet, the rover drives around taking photos and collecting rock and soil samples.

Several rovers have been sent to Mars. The most recent was Curiosity. It has 17 cameras, a small laser, a microscope, an X-ray spectrometer, and a long robotic arm. It also has six wheels, each with large treads, which is important on the sandy surface of Mars. It was designed this way after a previous rover got stuck. Curiosity arrived on Mars in 2012 and is still roving around the red planet, sending scientific information back to Earth.

GRACE
Computer
Programmer

1. How are rovers useful for studying objects in space?

_____

_____

_____

_____

2. What questions do you have about an aerospace engineer's work?

_____

_____

_____

_____

? Essential Question
How do the Sun, Earth, and the Moon interact?

_____

_____

_____

_____

⚙ Science and Engineering Practices

I will analyze and interpret data.

Like an aerospace engineer, you will analyze and interpret data about the movements of Earth and the Moon.

Inquiry Activity
Shadow Measurements

How does the Sun's position in the sky change during the day?

Make a Prediction Will the length and direction of a shadow change throughout the day? Explain.

_____

_____

_____

_____

Materials
☐ chalk
☐ ruler, meterstick, or tape measure
☐ graph paper

Carry Out an Investigation

BE CAREFUL You should not look directly at the Sun.

1. On a sunny morning, go outside to the location your teacher indicates.

2. Carefully shading your eyes, note the location of the Sun in relationship to a building, tree, or other landmark.

3. With a partner, find some space away from other students. Stand with your back to the Sun.

4. Have your partner make one chalk mark at the tip of your toes, and another mark at the very end of your shadow.

5. **Record Data** Measure the distance between the two chalk marks, and record it in the chart below.

6. Repeat these steps at midmorning, midday, and twice in the afternoon. Try to stand in the same place each time.

Time of Day				
Length of Shadow				

7 Analyze Data Use a separate sheet of graph paper to create a bar graph with length of shadow on the vertical axis and time of day on the horizontal axis. Remember to label your graph.

Communicate Information

1. How did the length of your shadow change throughout the day?

_____

_____

_____

2. Explain the pattern of change in the length of your shadow.

_____

_____

_____

_____

_____

_____

_____

_____

3. Describe the path of the Sun across the sky during the day.

_____

_____

_____

_____

Glue your graph here.

Obtain and Communicate Information

abc Vocabulary

Use these words when explaining movements of
the Sun, Earth, and the Moon.

orbit   gravity   inertia

revolution   rotation

Space

📖 Read pages 220–223 in the *Science Handbook*. Answer
the following questions after you have finished reading.

1. What is an orbit?

_____

_____

_____

2. How many times larger is the Sun compared to Earth?
   How does size relate to gravitational pull?

_____

_____

_____

3. Explain how gravity and inertia are involved with Earth's orbit.

_____

_____

_____

_____

4. Explain how the Moon's gravity affects Earth.

_____

_____

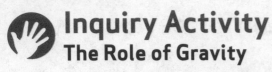

Inquiry Activity
The Role of Gravity

How does gravity affect the movement of objects in space? You will use a ball and string to simulate the role of gravity between the Sun and Earth.

Materials

☐ tennis ball

☐ cloth large enough to cover the ball, approximately 25 cm²

☐ 1.5 meters of string

Carry Out an Investigation

1. Wrap the cloth around the ball. Pull the corners of the cloth together and tie them in a knot.

2. Securely tie the string to the cloth at the knot.

3. Stand apart from other students, and slowly spin the ball in a circle.

4. On your teacher's signal, let go of the string. Be sure no students are in the way.

Communicate Information

5. While swinging the ball, what did you feel happening between the string and your hand?

_____

_____

6. How does this activity model the interaction between the Sun and Earth?

_____

_____

_____

7. What happened when you let go of the string?

_____

_____

8. What forces caused this to happen? Explain.

_____

_____

9. In the left box, draw a diagram of you swinging the ball in a circle.
Use arrows to indicate the directions of the two forces involved.
In the right box, draw a second diagram of Earth orbiting the Sun.
Use arrows to indicate the directions of the two forces involved.

What is Gravity?

Read *What is Gravity?* about the effects of the force of gravity.
Answer the following questions after you have finished reading.

10. When does the force of gravity between two objects decrease?

_____

_____

_____

11. What prevents planets from being pulled into the Sun?

_____

_____

_____

_____

Earth in Space

📖 Read pages 230–235 in the *Science Handbook*. Answer the following questions after you have finished reading.

12. Describe the difference between a rotation and a revolution.

_____

_____

_____

_____

_____

13. What causes the four different seasons in the Northern Hemisphere?

_____

_____

_____

_____

_____

_____

_____

_____

FOLDABLES®

Cut out the Notebook Foldables given to you by your teacher.
Glue the anchor tabs as shown below. Use what you have
learned to show how Earth's tilt on its axis causes the seasons.

Glue anchor tab here.

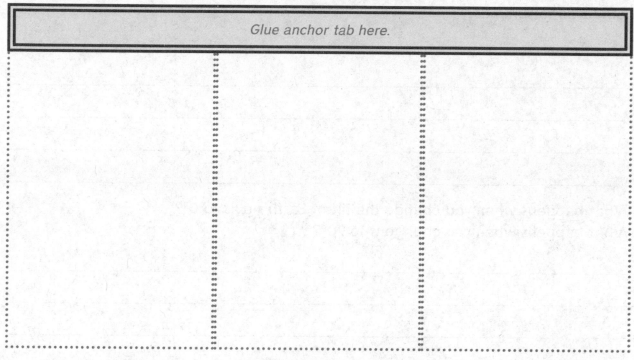

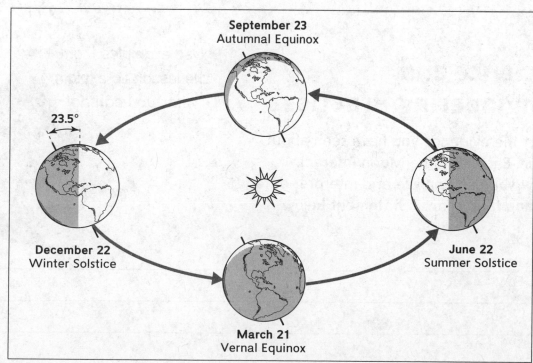

September 23
Autumnal Equinox

23.5°

December 22
Winter Solstice

June 22
Summer Solstice

March 21
Vernal Equinox

Earth Movements

▦ Investigate the interaction between Earth and the Sun as you conduct the simulation. Answer the questions after you have finished.

14. Observe the simulation without changing any of the settings. What is different about the angle of sunlight at noon in winter compared to summer?

_____

_____

_____

_____

15. What happens when you change the tilt of Earth's axis to 0°? What happens when you change it to 90°?

_____

_____

_____

_____

⚙ Science and Engineering Practices

Think about the evidence you have seen about how the Sun, Earth and the Moon interact. Express how you can analyze and interpret data by completing the "I can..." statement below.

Use examples from the lesson to explain what you can do!

I can _____

_____

_____

Research, Investigate, and Communicate
Midnight Sun and Polar Night

Research You will research the phenomena of midnight Sun and polar night and how they affect the polar regions.

Communicate Information

1. Describe the midnight Sun phenomenon.

_____

_____

_____

2. Where does midnight Sun occur?

_____

3. When is midnight Sun likely to occur in the Arctic? The Antarctic?

_____

_____

_____

4. How does midnight Sun affect people who live in these regions?

_____

_____

_____

5. Describe the polar night phenomenon.

_____

_____

_____

6. Where does polar night occur?

_____

7. When is polar night likely to occur in the Arctic? The Antarctic?

_____

_____

_____

_____

8. How does polar night affect people who live in these regions?

_____

_____

_____

Analyze Data Sketch the tilt of Earth relative to the Sun's rays when midnight Sun occurs. Then sketch the tilt of Earth when polar night occurs. Label the date and location on both of your sketches.

⚙ Performance Task
Three Cities

Research You will choose three different cities to research; one in the Northern Hemisphere, one in the Southern Hemisphere, and one near the Equator. Record the temperature and precipitation data for each city in the tables below and on the next page.

City 1 _____	Jan	Feb	Mar	Apr	May	Jun	Jul	Aug	Sep	Oct	Nov	Dec
Average High Temp (°C)												
Average Precipitation (cm)												

City 2 _____	Jan	Feb	Mar	Apr	May	Jun	Jul	Aug	Sep	Oct	Nov	Dec
Average High Temp (°C)												
Average Precipitation (cm)												

City 3 _____	Jan	Feb	Mar	Apr	May	Jun	Jul	Aug	Sep	Oct	Nov	Dec
Average High Temp (°C)												
Average Precipitation (cm)												

Analyze Data Use a separate sheet of graph paper to create line graphs that show how the climates of the three cities are similar and different. Use a different colored pencil to represent each city. Create a key for your graphs.

1. Analyze the data. Explain how the climate of each city is based on its location and the movement of Earth around the Sun.

_____

_____

_____

_____

Crosscutting Concepts
Communicate Information
Patterns

2. How does the timing of seasons compare in cities north and south of the Equator?

_____

_____

3. What relationship did you find among the temperatures in the three cities?

_____

_____

_____

? Essential Question

How do the Sun, Earth, and the Moon interact?

Think about the photos of midnight Sun and polar night you saw at the beginning of the lesson. How do those two phenomena show how the Sun, Earth, and the Moon interact?

_____

_____

_____

⚙ Science and Engineering Practices

Review the "I can . . ." statement you wrote earlier in the lesson. Explain what you have accomplished in this lesson by completing the "I did . . ." statement.

Now that you're done with the lesson, share what you did!

I did _____

_____

_____

Patterns of the Moon

PAGE KEELEY
SCIENCE
PROBES

Patterns of the Moon

There are regular, repeating patterns of the Moon that can be predicted. Put an X in any of the boxes that are examples of things that can be predicted by regular patterns of the Moon.

seasons	length of a day	lunar eclipses
solar eclipses	tides	moon phases
length of shadows	direction of shadows	sunrise
position of the Moon in the sky	sunset	weather

Explain your thinking. How did you decide which things could be predicted by regular patterns of the Moon?

_____

_____

_____

_____

_____

_____

Science in Our World

Look at the two photos of the Moon. What questions do you have about the changing appearance of the Moon?

_____

_____

_____

_____

_____

Read about a selenographer and answer the questions on the next page.

Selenographers study the surface of the Moon.

GRACE
Computer Programmer

STEM Career Connection
Selenographer

As a selenographer, I work for the National Aeronautics and Space Administration, also known as NASA. I work at the Lunar Science Institute at NASA's Goddard Flight Center in Maryland. I analyze data from various lunar orbiter and rover missions. Our goal is to have a complete and accurate map of the Moon's surface in order to prepare for future crewed missions to our nearest neighbor in space.

An accurate lunar map is very important to future Moon missions. A map of the Moon shows us where there are useful resources that future astronauts could utilize. Future missions could last weeks, months, or years. It would be much easier if astronauts didn't have to bring all of their supplies from Earth. For example, there may be water in the form of ice on the Moon. Water can be used for drinking, washing, and cooking. It can also be used to make breathable air and even rocket fuel!

1. Why is an accurate map of the Moon important for Moon missions?

_____

_____

_____

_____

2. How does a selenographer make accurate maps?

_____

_____

_____

_____

? Essential Question

What causes the repeating pattern of the Moon's appearance?

_____

_____

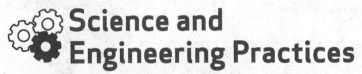 Science and Engineering Practices

I will engage in argument from evidence.

Like a selenographer, you will use evidence to support an argument of why the Moon seems to change shape.

Inquiry Activity
Moon Phases

How does the Moon's orbit affect its appearance?

Make a Prediction How does the Moon's apparent shape change during the month? Explain.

_____

_____

_____

_____

_____

Materials
☐ pencil
☐ drawing paper
☐ small foam ball that is half white and half black
☐ sharpened pencil
☐ circle cut from yellow construction paper
☐ tape

Carry Out an Investigation

BE CAREFUL Always use science materials appropriately.

1 Attach the "Sun" circle to the wall with tape.

2 Carefully insert the sharpened pencil into the foam ball on the dividing line between the white and black sides. The white side represents the half lighted by the Sun. The black half represents the dark side.

3 Have a classmate sit and represent Earth while you stand, holding up the pencil and ball that that represents the Moon.

4 Beginning directly between the Sun and your partner, hold the Moon so that the white half faces the Sun.

5 Your partner should observe the Moon and draw the shape that represents how much of the Moon's white side is visible.

6 Move in an arc about 45° to your partner's right, making sure that the white side of the Moon is still facing the Sun.

7 Your partner should again observe the Moon and draw the shape that represents how much of the Moon's white side is visible.

8 Continue moving around your partner, stopping every 45° for a drawing until you are back in front of the Sun. Make sure the white side is always pointed toward the Sun.

9 Switch places with your partner, and repeat steps 3-8.

Communicate Information

1. Are all of your drawings the same? Explain.

_____

_____

_____

2. Did the amount of sunlight reaching the Moon ever change?
   Explain.

_____

_____

3. If not, then why are your drawings different?

_____

_____

_____

4. **Construct an Explanation** Explain how this activity models
   the phases of the Moon.

_____

_____

_____

5. Compare your drawings to photos of the Moon's phases. How well
   do your drawings match the photos?

_____

_____

_____

_____

Obtain and Communicate Information

abc Vocabulary

Use these words when explaining the patterns of the moon.

satellite   phase   tide

solar eclipse   lunar eclipse

Earth's Moon

Read pages 236–237 in the *Science Handbook*. Answer the following questions after you have finished reading.

1. How long does it take the Moon to revolve around Earth?

_____

2. List and describe each of the Moon's phases.

_____

_____

_____

_____

_____

_____

_____

_____

_____

_____

_____

FOLDABLES®

Cut out the Notebook Foldables given to you by your teacher.
Glue the anchor tabs as shown below. Use what you have learned
in the lesson to complete the activity.

Glue anchor tab here

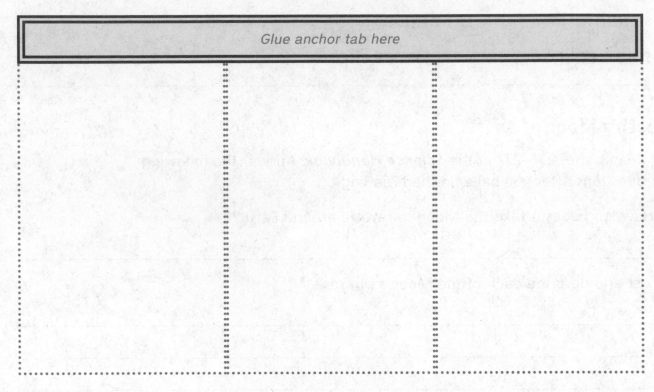

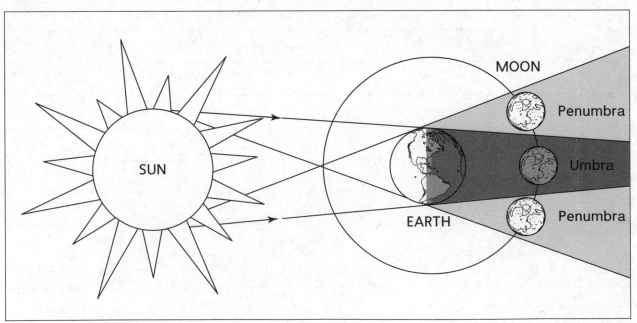

Tides and Eclipses

👁 Read *Tides and Eclipses* about the causes of these events between Earth and the Moon. Answer the following questions after you have finished reading.

3. What causes tides?

_____

_____

4. Why are there two high and two low tides every day?

_____

_____

_____

_____

_____

_____

5. Why do the Sun and the Moon both contribute to tides?

_____

_____

6. When does a solar eclipse occur?

_____

_____

7. When does a lunar eclipse occur?

_____

_____

_____

_____

Tide Research

Research You will choose a coastal town and research its high and low tides. Use what you know about the movements of the Moon to analyze patterns in the data.

1 Choose a coastal town.

_____

2 Use the Internet to find a tide table that shows the town's high and low tides.

3 **Record Data** Record the high and low tides for a one-week period.

Day/Date	1st Tide	2nd Tide	3rd Tide	4th Tide

4 Make a line graph of time and the height of high tides on a separate sheet of graph paper. Label your graph.

5 Make a line graph of time and the height of low tides on a separate sheet of graph paper. Label your graph.

8. Analyze Data About how much time passed between two high tides?

_____

_____

9. Were all of the high tides the same height?

_____

_____

Glue your graph here.

10. What is the average height difference between the high tides and low tides?

_____

_____

_____

11. Explain how movements of the Moon affect life on Earth.

_____

_____

_____

_____

Science and Engineering Practices

Think about what you have learned about how the Moon's movements can affect Earth. Express how you can analyze data and argue from evidence by completing the "I can..." statement below.

Use examples from the lesson to explain what you can do!

I can _____

_____

_____

_____

Research, Investigate, and Communicate

Other Moons

A *satellite* is an object in space that orbits another object.
A *moon* is a natural satellite that orbits a planet.

Research Choose a planet other than Mercury, Venus, or Earth.
Use reliable resources to find information about the moons that orbit
the planet you chose. Record the following information.

1. Planet

_____

2. How many moons does the planet have? List the names or some
   of the names of the moons.

_____

_____

3. What is the diameter of this planet's largest moon?

_____

4. Earth's moon has a diameter of 3,475 kilometers (km).
   Which is larger: Earth's moon or the moon(s) you researched?

_____

_____

5. How long does it take for the moon(s) to orbit the planet?

_____

_____

_____

6. List some interesting facts that you have learned about the moon(s).

_____

_____

_____

Performance Task
Phases of the Moon

You will create a model of the Moon's phases. Use your drawings from earlier in the lesson to help you.

Make a Model

BE CAREFUL Wear safety goggles. Use caution when using scissors.

1. Cut an 8 centimeter (cm) circle from blue paper. Glue it in the center of a sheet of construction paper (not black or white). Label the circle "Earth."

2. Choose one edge of the paper to be the Sun. Glue a thin strip of yellow construction paper on that edge.

3. Cut out eight 3-cm circles from white construction paper and four 3-cm circles from black construction paper. Cut each of the black circles in half.

4. Glue a black half-circle onto each full white circle.

5. Arrange the little circles around Earth so that the white half of each is closer to the Sun and the black half is facing away from the Sun.

6. Based on your drawings and any materials your teacher approves, create a 5-cm circle that shows what phase of the Moon is visible from Earth at each of the Moon's locations in orbit.

Materials

- [ ] colored construction paper
- [ ] markers
- [ ] glue
- [ ] safety goggles
- [ ] scissors
- [ ] ruler

Think like a selenographer and model the phases of the Moon.

Communicate Information

⚙ Crosscutting Concepts
Patterns

1. **Construct an Explanation** Why does the pattern of phases repeat every 29 days?

_____

_____

_____

2. Does the amount of sunlight that reaches the Moon ever change? Explain.

_____

_____

_____

3. Does your location on Earth determine which phases of the Moon you see? Explain.

_____

_____

_____

4. If you were on the Moon, would you see phases of Earth? Explain.

_____

_____

_____

_____

_____

? Essential Question

What causes the repeating pattern of the Moon's appearance?

Think about the photos of the Moon at night and during the day from the beginning of the lesson. What causes the repeating patterns of the Moon's appearance?

_____

_____

_____

_____

⚙ Science and Engineering Practices

Review the "I can . . ." statement you wrote earlier in the lesson. Explain what you have accomplished in this lesson by completing the "I did . . ." statement.

Now that you're done with the lesson, share what you did!

I did _____

_____

_____

_____

Objects in Space

PAGE KEELEY
SCIENCE
PROBES

Earth's Gravity

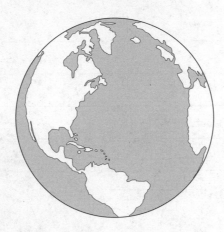

Four friends were talking about gravity. They each had different ideas. This is what they said:

Angie: I think Earth's gravity only pulls on things that are close to Earth's surface.

Tamara: I think Earth's gravity only pulls on things that are close to the Earth and in Earth's atmosphere.

Carlos: I think Earth's gravity pulls on things close to the Earth, including some things in space.

Mi-Ling: I think Earth's gravity pulls everything in the solar system toward it.

Who do you agree with most? _____

Explain why you agree.

_____

_____

_____

_____

_____

Science in Our World

Look at the photo of a meteor shower. What are meteors? What questions do you have about what is happening?

_____

_____

_____

_____

Read about an astronaut and answer the questions on the next page.

STEM Career Connection

Astronaut

Today is my first day training to be an astronaut for NASA. I have been preparing for this for years, earning my degree in physical science while also studying space technology, and space science. I have even had medical and scuba training. I am also in the process of learning several foreign languages so that I can communicate with astronauts from other countries. It sounds like a lot to manage, but I know it will be rewarding to experience outer space.

I am hoping to be assigned to an upcoming mission to the International Space Station, or ISS. In order to be selected, I will have to study all of the complex controls of this amazing orbiting lab. This will involve participating in flying and microgravity simulations as well as preparing for the work to be completed once I arrive on the ISS.

Astronauts train to live and work in space.

GRACE
computer programmer

Name _____ Date _____

1. What kinds of training does an astronaut need to have?

_____

_____

_____

_____

2. What questions do you have about human spaceflight?

_____

_____

_____

? Essential Question
What other objects can be found in space?

_____

_____

⚙ Science and Engineering Practices

I will engage in argument from evidence.

Like an astronaut, you will use evidence to explain the characteristics of objects in space.

Inquiry Activity
Modeling Moon Craters

What factors affect the size of craters that form when speeding objects strike the surface of the Moon?

Make a Prediction How does the size of an object affect the size of the crater it forms when it strikes the Moon?

_____

_____

_____

Materials

☐ safety goggles

☐ newspaper

☐ shallow pan

☐ sand, flour, or fine dirt

☐ different sized marbles

☐ plastic spoon

☐ ruler

Carry Out an Investigation

BE CAREFUL Wear safety goggles.

1 Cover the floor with newspaper, and place the pan on the newspaper.

2 Fill the pan with sand or flour to about 2 centimeters (cm) deep.

3 Drop each of the marbles from the same height into a different area of the pan.

4 **Record Data** Carefully remove each marble with the plastic spoon and measure the diameter of each crater and record it in the table.

Size of Marble	Diameter of Crater Formed

Communicate Information

Analyze Data Answer the questions based on the data you collected.

1. What did you see at the crater sites? Why did this happen?

_____

_____

_____

2. How does the size of the crater compare to the size of the marble?

_____

_____

3. How does this model represent what happens when an object hits the surface of the Moon?

_____

_____

_____

⚙ Crosscutting Concepts
Cause and Effect

4. Consider how the movement of objects in space could lead to impact craters here on Earth. If our closest neighbor, the Moon, has impact craters, what does that tell you about the likelihood of objects from space impacting Earth?

_____

_____

5. Many of the Moon's craters were created long ago. Since there is no erosion on the Moon to destroy the craters, there is a near perfect record of the impacts. Why aren't craters as visible here on Earth?

_____

_____

Obtain and Communicate Information

🔤 Vocabulary

> **Use these words when explaining objects in space.**
>
> asteroid   meteor   comet

The Solar System

📖 Read pages 222 and 224–227 in the *Science Handbook.* Answer the following questions after you have finished reading.

1. Why do the planets orbit the Sun?

_____

_____

2. Name and describe one of the four inner planets.

_____

_____

_____

3. What are asteroids?

_____

_____

_____

4. Name and describe one of the four outer planets.

_____

_____

FOLDABLES®

Cut out the Notebook Foldables given to you by your teacher.
Glue the anchor tabs as shown below. Use what you have learned
to describe and define the objects in space.

Glue anchor tab here

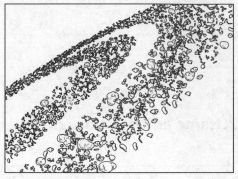

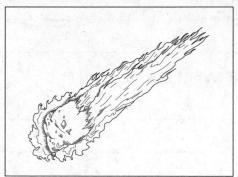

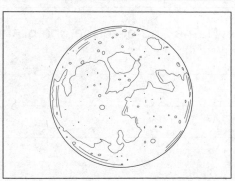

Other Objects in the Solar System

📖 Read pages 228–229 in the *Science Handbook.* Answer the following questions after you have finished reading.

6. What is the difference between a meteoroid, a meteor, and a meteorite?

_____

_____

_____

_____

7. What is a comet?

_____

_____

_____

8. Describe what happens to a comet as it gets closer to the Sun.

_____

_____

_____

_____

Super Space Objects

📖 Use your *Science Handbook* and other resources to research other objects in space.

9. In what area of the solar system do comets originate?

_____

_____

_____

10. What causes comets to move through the solar system?

_____

_____

_____

_____

11. In the space provided, draw a diagram of a comet. Label at least three specific parts of a comet.

12. Name four of the largest asteroids.

_____

13. How do scientists think the asteroid belt formed?

_____

_____

_____

14. Why are meteors sometimes called shooting stars?

_____

_____

_____

_____

15. What is the difference in size between asteroids and meteoroids?

_____

_____

_____

16. What serious effect can a meteorite have?

_____

_____

_____

17. How do the movements of comets, meteors, and asteroids compare to the movements of the planets?

_____

_____

_____

_____

_____

_____

⚙ Science and Engineering Practices

Think about what you have learned about how objects in space move. Tell how you can use what you have learned as evidence by completing the "I can . . ." statement below.

I can _____

_____

Use examples from the lesson to explain what you can do!

Name _____ Date _____

Research, Investigate, and Communicate

Comets, Asteroids, and Meteors

Read the characteristics listed below. Add each letter to the appropriate location on the Venn diagram.

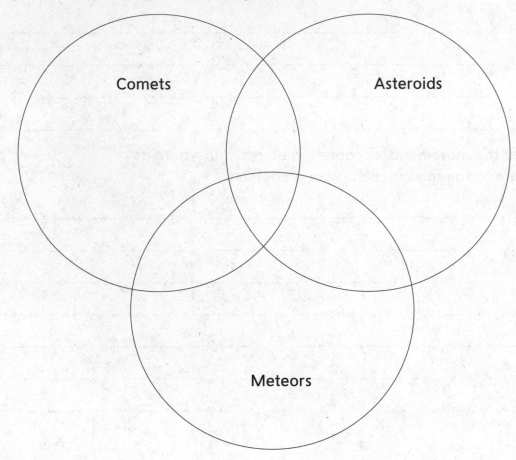

A. Progress across the sky very slowly over weeks or months

B. Remnants of the formation of the solar system

C. Reflect sunlight

D. Mostly rocky composition

E. Made mostly of ice

F. Orbit the Sun in highly elliptical orbits

G. Most are found between Mars and Jupiter

H. Most are found beyond Neptune

I. Can measure several km in diameter

J. Most are less than 1 km in diameter

K. Most are less than 100 m in diameter

L. Most burn up as they enter Earth's atmosphere

M. Streak across the sky very quickly

N. Tail always points away from the Sun

Performance Task
Model the Solar System

You are going to use a model of the solar system to explain how the size and location of each object affects its force of gravity on other objects.

Make a Model

BE CAREFUL Wear safety goggles. Use caution with the scissors.

1 Cut circles from construction paper to represent the Sun, the inner planets, the asteroid belt, and the outer planets.

2 Lay the string out in a straight line on the floor. Put the Sun at one end.

3 In the table below, astronomical units (AU) have been converted to centimeters. Use these measurements to place each space object at the correct distance from the Sun on the string.

Space Object	Distance from the Sun (cm)
Mercury	4 cm
Venus	7 cm
Earth	10 cm
Mars	15 cm
Asteroid Belt	28 cm
Jupiter	52 cm
Saturn	96 cm
Uranus	192 cm
Neptune	300 cm

Materials

- [ ] safety goggles
- [ ] scissors
- [ ] construction paper
- [ ] 5 meters of string
- [ ] meterstick

Communicate Information

1. How does distance from the Sun affect the properties of the planets? Give an example to explain your answer.

_____

_____

_____

_____

_____

2. Compare the physical properties of the smaller, inner planets and the outer gas giants.

_____

_____

_____

_____

_____

3. Describe how the forces of gravity and inertia affect space objects in our solar system.

_____

_____

_____

_____

_____

? Essential Question
What other objects can be found in space?

Think about the photo of a meteor shower at the beginning of the lesson. Use what you have learned to name other objects that can be found in space.

_____

_____

_____

_____

⚙ Science and Engineering Practices

Review the "I can . . ." statement you wrote earlier in the lesson. Explain what you have accomplished in this lesson by completing the "I did . . ." statement.

Now that you're done with the lesson, share what you did!

I did _____

_____

_____

_____

_____

Name _____   Date _____

Stars and Star Patterns

 PAGE KEELEY
**SCIENCE
PROBES**

Constellations

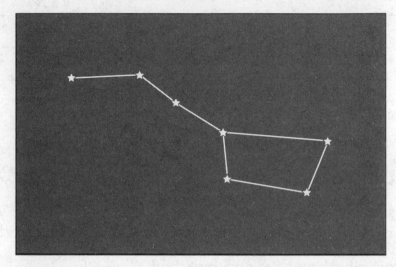

Kendra and Jake were looking at the night sky. They recognized a constellation called The Big Dipper. They each had different ideas about the constellations. This is what they said:

Jake: *The stars form the same constellations throughout the year. They keep the same patterns even though we see different constellations at different times of the year.*

Kendra: *The stars form different constellations throughout the year. They are always changing their patterns, which is why we see different constellations at different times of the year.*

Who do you agree with more? _____

Explain why you agree.

_____

_____

_____

_____

_____

Science in Our World

▶ Watch the video of a black hole. What questions do you have about black holes?

_____

_____

_____

Read about an astronomer and answer the questions on the next page.

Astronomers use telescopes to observe the universe.

STEM Career Connection
Astronomer

As an astronomer, I work at the Very Large Array, a collection of radio telescopes at the National Radio Astronomy Observatory in New Mexico. My job is to investigate how galaxies form and to determine if supermassive black holes exist at the centers of distant galaxies.

Other astronomers use visual light telescopes that use lenses or mirrors to bend light from distant objects. Radio astronomy uses radio telescopes to detect very faint signals given off by stars and galaxies. By analyzing the signals, we can tell a lot about a star or galaxy. Since radio telescopes do not depend upon visible light, they can observe the universe during daytime or nighttime and in any type of weather.

GRACE
Computer Programmer

1. What is special about a radio telescope?

_____

_____

_____

2. What questions do you have about the astronomer's work?

_____

_____

? Essential Question
What are stars, and why are some brighter than others?

_____

_____

_____

_____

_____

Science and Engineering Practices

I will engage in argument from evidence.

Think like an astronomer and use evidence to explain differences in the brightness of stars.

Inquiry Activity
Star Brightness

What factors affect how bright a star looks from Earth?

Make a Prediction How does a star's distance from Earth affect its apparent brightness?

_____

_____

Materials

☐ one large flashlight

☐ one pocket flashlight

☐ meterstick

Carry Out an Investigation

1. Place a piece of tape on the floor at a distance of five steps from the wall.

2. Shine a flashlight or "star" on the wall from this spot. Observe how bright the circle looks. Have a partner measure the diameter of the circle of light.

3. What will happen to the size and apparent brightness of the circle of light if you move the star to 3 steps from the wall? What about 1 step from the wall? Test your predictions.

1. Record Data Did the amount of light put out by the star change? Explain.

_____

_____

_____

2. Suppose you keep the star five steps from the wall, but this time hold two flashlights, one the same as before and one smaller flashlight. How does the light on the wall compare?

_____

_____

3. What could you do to get the two "stars" to have the same size circle of light on the wall?

_____

_____

_____

Communicate Information

4. Explain how this activity models different degrees of brigtness in stars.

_____

_____

_____

_____

5. If you could see only the wall and not the flashlights, how would you know if they were equally bright lights at the same distance or two unequal lights at different distances?

_____

_____

6. If you saw two unequally bright spots on the wall and could not see the flashlights, how would you determine the quality or distance of the light?

_____

_____

7. Draw two diagrams of your experiment. In one diagram, show the two flashlights the same distance from the wall. In the other diagram, show the flashlights at different distances from the wall. Label how the brightness of the lights compares in each diagram.

Obtain and Communicate Information

🔤 Vocabulary

Use these words when explaining stars and star brightness.

star   light-year   constellation

nebula   white dwarf   supernova

black hole

Stars

📖 Read pages 238–239 in the *Science Handbook*. Answer
the following questions after you have finished reading.

1. What is a star?

_____

_____

2. What does the color of a star tell us?

_____

_____

_____

3. How old is the Sun? How much longer will it last?

_____

_____

4. What is a light-year? What do light-years tell us?

_____

_____

_____

Constellations

📖 Read pages 244–245 in the *Science Handbook.* Answer the following questions after you have finished reading.

5. What is a constellation?

_____

_____

_____

6. How many known constellations are there? _____

7. How can recognizing these star patterns be useful?

_____

_____

_____

8. How does Earth's rotation and revolution affect which stars and constellations we see?

_____

_____

_____

_____

_____

🔎 Explore the Digital Interactive *Constellations* about patterns of stars in the night sky. Answer the questions after you have finished.

9. Where and when is the constellation of Orion most visible?

_____

_____

10. Where and when is the constellation of Pegasus most visible?

_____

_____

FOLDABLES

Cut out the Notebook Foldables given to you by your teacher.
Glue the anchor tabs as shown below. Use what you have learned
about the terms to complete the activity.

Glue anchor tab here.

Glue anchor tab here.

Our Star, the Sun

📖 Read pages 246–247 in the *Science Handbook.* Answer the following questions after you have finished reading.

⚙️ Crosscutting Concepts
Scale, Proportion, and Quantity

11. What is the Sun mostly made of?

_____

12. How big is the Sun? How much of the solar system's mass is in the Sun?

_____

_____

_____

13. What is the temperature on the surface of the Sun? What is the temperature at the Sun's core?

_____

14. How long does it take heat energy to travel from the Sun's core to its surface?

_____

⚙️ Science and Engineering Practices

Think about what have you learned about the Sun and other stars. Tell how you can argue from evidence by completing the "I can . . ." statement below.

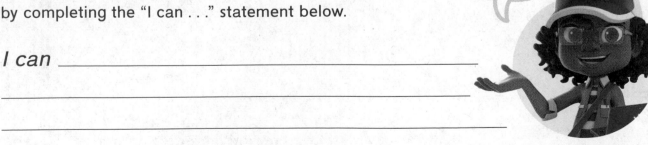

Use examples from the lesson to explain what you can do!

I can _____

_____

_____

Star Cycles

▶ Watch the video *Star Cycles* on how stars are formed, live, and eventually die.

📖 Then, read pages 240–243 in the *Science Handbook.* Answer the following questions after you have finished reading.

1. What does a star's cycle depend on?

_____

_____

2. When does a star's cycle end?

_____

_____

3. Describe how a nebula turns into a protostar.

_____

_____

_____

_____

_____

_____

4. What stages will a star like the Sun go through?

_____

_____

_____

5. What eventually happens to larger stars?

_____

_____

6. What happens to massive stars at the end of their life cycles?

_____

_____

Performance Task
Model a Constellation

Materials

☐ small, white foam balls

☐ craft sticks or straws

☐ black construction paper

You will choose a constellation to research. Use what you have learned about stars and their brightness to explain the pattern of the constellation.

1 List the stars in the constellation and their distances from Earth. Note where and when in the night sky your constellation is visible.

_____

_____

_____

_____

2 Sketch your constellation. Label the stars that have names.

3 Build a model of your constellation. Use white foam balls for the stars, and connect them with craft sticks or straws. Follow the pattern in your drawing.

Communicate Information

1. How can you use your model to teach other students about stars and star patterns?

_____

_____

_____

❓ Essential Question
What are stars and why are some brighter than others?

▶ Think about the black hole video from the beginning of the lesson. Use what you have learned to explain how the brightness of stars can vary.

_____

_____

_____

_____

_____

⚙ Science and Engineering Practices

Review the "I can . . ." statement you wrote earlier in the lesson. Explain what you have accomplished in this lesson by completing the "I did . . ." statement.

> Now that you're done with the lesson, share what you did!

I did _____

_____

_____

_____

The Solar System and Beyond

⚙ Performance Project
Modeling Space Objects

▦ Revisit the simulation *Earth Movements*. Earth's orbit, the Moon's orbit around Earth, and Earth's rotation about its axis result in observable patterns. These include day and night; daily changes in the length and direction of shadows; and different positions of the Sun, the Moon, and stars at different times of the day, month, and year. Choose one of these patterns and explain on the lines below how using the simulation helped you to understand the pattern.

_____

_____

_____

_____

_____

_____

_____

_____

_____

_____

_____

_____

_____

_____

How can a space scientist analyze data to show how space objects interact?

Use what you have learned about space objects to create a graphic display that illustrates your understanding of one of the following patterns in our solar system:

1. changes in the length and direction of shadows

2. day and night

3. seasonal changes to stars and constellations in the night sky

Make a sketch of your display in the box below. Use the sketch to make your display. Compare it with your classmates' displays.

 Explore More in Our World

Did you learn the answers to all of your questions from the beginning of the module? If not, how could you design an experiment or conduct research to help answer them?

An Interview with
Dinah Zike Explaining
Visual Kinesthetic Vocabulary®, or VKVs®

What are VKVs and who needs them?

" VKVs are flashcards that animate words by kinesthetically focusing on their structure, use, and meaning. VKVs are beneficial not only to students learning the specialized vocabulary of a content area, but also to students learning the vocabulary of a second language. "

Dinah Zike | Educational Consultant
Dinah-Might Activities, Inc. – San Antonio, Texas

Why did you invent VKVs?

" Twenty years ago, I began designing flashcards that would accomplish the same thing with academic vocabulary and cognates that Foldables® do with general information, concepts, and ideas—make them a visual, kinesthetic, and memorable experience. "

I had three goals in mind:

- **Making two-dimensional flashcards three-dimensional**

- **Designing flashcards that allow one or more parts of a word or phrase to be manipulated and changed to form numerous terms based upon a commonality**

- **Using one sheet or strip of paper to make purposefully shaped flashcards that were neither glued nor stapled, but could be folded to the same height, making them easy to stack and store**

Why are VKVs important in today's classroom?

" At the beginning of this century, research and reports indicated the importance of vocabulary to overall academic achievement. This research resulted in a more comprehensive teaching of academic vocabulary and a focus on the use of cognates to help students learn a second language. Teachers know the importance of using a variety of strategies to teach vocabulary to a diverse population of students. VKVs function as one of those strategies. "

Dinah Zike Explaining
Visual Kinesthetic Vocabulary®, or VKVs®

How are VKVs used to teach content vocabulary to EL students?

" VKVs can be used to show the similarities between cognates in Spanish and English. For example, by folding and unfolding specially designed VKVs, students can experience English terms in one color and Spanish in a second color on the same flashcard while noting the similarities in their roots. "

How are VKVs used to teach content vocabulary?

" As an example, let's look at content terms based upon the combining form *–vore*. Within a unit of study, students might use a VKV to kinesthetically and visually interact with the terms *herbivore, carnivore,* and *omnivore*. Students note that *–vore* is common to all three words and it means "one that eats" meat, plants, or both depending on the root word that precedes it on the VKV. When the term *insectivore* is introduced in a classroom discussion, students have a foundation for understanding the term based upon their VKV experiences. And hopefully, if students encounter the term *frugivore* at some point in their future, they will still relate the *–vore* to diet, and possibly use the context of the word's use to determine it relates to a diet of fruit. "

Dinah Zike's book Foldables, Notebook Foldables, & VKVs for Spelling and Vocabulary 4th-12th won a Teachers' Choice Award in 2011 for "instructional value, ease of use, quality, and innovation"; it has become a popular methods resource for teaching and learning vocabulary.

Dinah Zike's
Visual
Kinesthetic
Vocabulary®

melting point

_____ is the temperature at which a substance changes state from a liquid into a gas.

_____ is the temperature at which a substance changes state from a liquid to a solid.

_____ is the particular temperature at which a substance changes state from a solid into a liquid.

✂ cut on all dashed lines   fold on all solid lines

Memory Maker: In each of the boxes below, draw a picture to define the term.

Freezing point

Melting point

Boiling point

freezing

boiling

Dinah Zike's
**Visual
Kinesthetic
Vocabulary**®

✂ cut on all dashed lines   fold on all solid lines

A _____ is a substance that has a high pH level and forms salts in chemical reactions with an acid.

14

_____ is a mixture of substances that are blended so completely that the mixture looks the same everywhere.

A substance's _____ is the maximum amount of the substance that can be dissolved by another substance.

(neutral)

7

solubility

A _____ is a substance that has a low pH level and forms salts in chemical reactions with a base.

1

Dinah Zike's
Visual
Kinesthetic
Vocabulary®

✂ cut on all dashed lines   ▭ fold on all solid lines

base

tion

acid

Memory Maker: Are **acid** and **base** antonyms or synonyms? Explain your answer. _____

Memory Maker: Both **solution** and **solubility** share the word part **solu-**. This word part is a Latin root that means "loosen." How does knowing the meaning of **solu-** help you remember the meanings of **solution** and **solubility**?

Dinah Zike's
Visual
Kinesthetic
Vocabulary®

Plant and Animal Needs

✂ cut on all dashed lines   ⬚ fold on all solid lines

photosynthesis

_____ is a green chemical in plant cells that allows plants to use the Sun's energy to make food.

A _____ is part of a plant cell that contains chlorophyll; where photosynthesis takes place in the cell.

chloroplast

Dinah Zike's
VKV Visual
Kinesthetic
Vocabulary®

✂ cut on all dashed lines   ▢ fold on all solid lines

is the food-making process in green plants that uses sunlight.

phyll

Memory Maker: The words **chlorophyll** and **chloroplast** share a word part: **chloro-**. This word part means "green." How does knowing the meaning of **chloro-** help you remember that **chlorophyll** and **chloroplasts** are found in plants?

Memory Maker: The word part **phot-** in **photo** and **photosynthesis** means "light." What does "light" have to do with the meaning of **photosynthesis**?

Dinah Zike's
Visual Kinesthetic Vocabulary®

✂ cut on all dashed lines

⬜ fold on all solid lines

cellular respiration

___ is the process of using oxygen to break down food into energy.

___ is the process of breaking down food into energy without using oxygen.

___ is the process of releasing energy from food molecules, such as glucose, which takes place in the mitochondria of a cell.

Dinah Zike's
Visual Kinesthetic Vocabulary®

✂ cut on all dashed lines

fold on all solid lines

Memory Maker: In **cellular respiration**, cells release energy. How is **cellular respiration** different from **aerobic respiration and anaerobic respiration?**

anaerobic

aerobic

Dinah Zike's
Visual
Kinesthetic
Vocabulary®

cut on all dashed lines

fold on all solid lines

water cycle

The _____ is the continuous trapping of nitrogen gas into compounds in the soil and its return to the air.

The _____ is the continuous exchange of carbon dioxide and oxygen among living things.

The _____ is the continuous movement of water between Earth's surface and the air, changing from liquid into gas into liquid.

Dinah Zike's
Visual
Kinesthetic
Vocabulary®

VKV

Matter in Ecosystems

✂ cut on all dashed lines   ⬜ fold on all solid lines

Memory Maker: Cycles are often represented with circles because, like circles, cycles are continuous. For each of the terms on this card, draw a circle diagram to define the term.

oxygen-
carbon dioxide

nitrogen

Dinah Zike's
Visual
Kinesthetic
Vocabulary®

✂ cut on all dashed lines

📩 fold on all solid lines

A _____ is the overlapping food chains in an ecosystem.

A _____ is the path that energy and nutrients follow in an ecosystem.

biotic factors

food chain

Dinah Zike's
Visual Kinesthetic Vocabulary®

VKV

✂ cut on all dashed lines

fold on all solid lines

nonliving parts of that ecosystem.

_____ are the effects on the ecosystem that are a result of the

_____ are living things in an ecosystem, such as plants, animals, or bacteria.

Memory Maker: The prefix **a-** is a word part that sometimes means "not." If **biotic** means "related to living organisms," then what does **abiotic** mean?

web

a

Memory Maker: In your own words, explain the difference between a **food chain** and a **food web**.

Dinah Zike's
Visual
Kinesthetic
Vocabulary®

✂ cut on all dashed lines   fold on all solid lines

mutualism

is a relationship in which one organism lives in or on another organism and benefits from that relationship while the host organism is harmed by it.

is a relationship between two kinds of organisms that benefits one without harming the other.

is a relationship between two kinds of organisms that benefits both.

Dinah Zike's
Visual
Kinesthetic
Vocabulary®

VKV

✂ cut on all dashed lines

▭ fold on all solid lines

Memory Maker: **Parasitism, commensalism, and mutualism** have the word part -ism in common. Circle the meaning of -ism.

relationship

kinds

harm

without

commensal

parasit

✂ cut on all dashed lines   ▭ fold on all solid lines

Dinah Zike's
**Visual
Kinesthetic
Vocabulary** ®

hydrosphere

The _____ is the part of Earth in which living things exist and interact.

The _____ is the layers of rock, dirt, and soil on Earth, including the mantle, cores, and crust.

The _____ is Earth's water, whether found on land or in oceans, including the fresh water in ice, lakes, rivers, and underground.

Dinah Zike's **Visual Kinesthetic Vocabulary** ®

✂ cut on all dashed lines ▢ fold on all solid lines

Memory Maker: The word part **sphere** is a Greek root that means "ball." What does a ball shape have to do with the meanings of **biosphere, geosphere,** and **hydrosphere?**

_____

geo

bio

✂ cut on all dashed lines   ▭ fold on all solid lines

precipitation

is the slow changing of a liquid into a gas when particles vaporize at the water's surface.

is the changing of a gas into a liquid as heat is removed.

is water that falls from clouds to the ground in the form of rain, sleet, hail, or snow.

Dinah Zike's
VKV
Visual
Kinesthetic
Vocabulary®

cut on all dashed lines   fold on all solid lines

Memory Maker: Use your own words to describe how **evaporation** and **condensation** lead to **precipitation**.

condensa

evapora

✂ --- cut on all dashed lines   📄 fold on all solid lines

_____ is Earth's solid, rocky surface.

_____ is the layer of Earth beneath the crust.

_____ is the central part of Earth.

Earth's core

cut on all dashed lines   fold on all solid lines

mantle

Memory Maker: Draw a picture that defines the terms **Earth's crust**, **Earth's mantle**, and **Earth's core**.

crust

Dinah Zike's
Visual
Kinesthetic
Vocabulary®

✂ cut on all dashed lines   fold on all solid lines

_____ is the
process of removing salt
from salt water in order
to be used by living
things.

saline

plain

renewable resource

Circle the word that
best describes a plain.

hilly

flat

mountainous

A _____
is a resource that
can be replanted
or replaced
naturally in a
short period of
time.

A _____ is
land near a river that
is likely to be under
water during a flood.

A _____
is a resource
that cannot be
replaced within
a short period of
time or at all.

Dinah Zike's
VKV
Visual
Kinesthetic
Vocabulary®

✂ cut on all dashed lines   ▭ fold on all solid lines

ation

Memory Maker: The word part **sal-** is a Latin root that means "salt." The word part **de-** is a prefix that means "opposite." The word part **-ation** is a suffix that means "process of." Do you think **saline** is what remains after the **desalination** of water? Explain your answer.

_____

Memory Maker: The term **floodplain** is a compound word—it is formed by joining the words **flood** and **plain**. How do the definitions of **flood** and **plain** help you understand what a **floodplain** is?

_____

Memory Maker: The word part **non-** is a prefix that means "not." The word part **re-** is a prefix that means "again." The word part **-able** is a suffix that means "able to be." Write the word that remains when **non-**, **re-**, and **-able** are removed from **nonrenewable** and **renewable**. Then use your own words to write definitions for **renewable** and **nonrenewable**.

_____

non

flood

de

Dinah Zike's
Visual
Kinesthetic
Vocabulary®

✂ cut on all dashed lines   fold on all solid lines

revolution

solar eclipse

A _____ is
a complete spin on an axis.

A _____ is
one complete trip of one object
around another object.

A _____ is a situation that
occurs when Earth, the Sun, and the Moon are in
a straight line and Earth's shadow falls across the
Moon.

A _____ is a blocking of
the Sun's light that happens when Earth passes
through the Moon's shadow.

Dinah Zike's
Visual Kinesthetic Vocabulary®

cut on all dashed lines

fold on all solid lines

Memory Maker: How are the meanings of **rotation** and **revolution** similar? How are they different? _____

Memory Maker: In each of the boxes below, draw a picture to define the term.

Lunar eclipse

Solar eclipse

lunar

rota